IS-800.b - National Response Framework,
An Introduction

By FEMA

Based on Public Domain Text

Course: **IS-800.B - National Response Framework, An Introduction**
Lesson: **1 - Overview**

Course Overview

This course provides an introduction to the National Response Framework.

At the end of this course, you will be able to describe:

- The purpose of the National Response Framework.

- The response doctrine established by the National Response Framework.

- The roles and responsibilities of response partners.

- The actions that support national response.

- The response organizations used for multiagency coordination.

- The relationship between planning and national preparedness.

Introducing the National Response Framework: Video Transcript

In recent years, our Nation has faced an unprecedented series of disasters and emergencies. As a result, our national response structures have evolved and improved to meet these threats. The National Response Framework is the next step in this evolution, and as such defines how we respond as a Nation. Based on best practices and stakeholder input, the Framework presents the guiding principles that enable all response partners to prepare for and provide a unified national response to disasters and emergencies – from the smallest incident to the largest catastrophe.

Building on the National Incident Management System, the Framework's coordinating structures align key roles and responsibilities fostering response partnerships at all levels of government, and with nongovernmental organizations and the private sector. Given its flexibility and scalability, the National Response

Framework is always in effect and elements can be implemented at any level and at any time.

The Framework establishes a response vision through five key principles. Let's take a closer look at these principles, starting with engaged partnership.

Engaged partnership means that leaders at all levels develop shared response goals and align capabilities so that no one is overwhelmed in times of crisis.

The next principle is tiered response. Incidents must be managed at the lowest possible jurisdictional level and supported by additional capabilities when needed.

The third principle is scalable, flexible, and adaptable operational capabilities. As incidents change in size, scope, and complexity, the response must adapt to meet requirements.

The fourth principle is unity of effort through unified command. Unity of effort respects the chain of command of each participating organization while harnessing seamless coordination across jurisdictions in support of common objectives.

The last principle is readiness to act. It is our collective duty to provide the best response possible. From individuals, households, and communities to local, tribal, State, and Federal governments, national response depends on our readiness to act.

The National Response Framework strives to improved coordination among all response partners. And through these partnerships, we can work together to help save lives and protect America's communities.

National Response Framework Purpose

The purpose of the National Response Framework is to ensure that all response partners across the Nation understand domestic incident response roles, responsibilities, and relationships in order to respond more effectively to any type of incident.

The Framework is written especially for government executives, private-sector and nongovernmental organization leaders, and emergency management practitioners.

National Response Framework Scope: Domestic Incident Response

The Framework provides structures for implementing national-level policy and operational coordination for domestic incident response.

The term "response" as used in this Framework includes:

- Immediate actions to save lives, protect property and the environment, and meet basic human needs.

- The execution of emergency plans and actions to support short-term recovery.

Why Is the Framework Always In Effect?

It is not always obvious at the outset whether a seemingly minor event might be the initial phase of a larger, rapidly growing threat.

The National Response Framework allows for the rapid acceleration of response efforts without the need for a formal trigger mechanism.

Key Concept: The Framework is always in effect, and elements can be implemented as needed on a flexible, scalable basis to improve response.

Part of a Broader Strategy

The National Response Framework is required by, and integrates under, a larger National Strategy for Homeland Security that:

- Serves to guide, organize, and unify our Nation's homeland security efforts.

- Reflects our increased understanding of the threats confronting the United States.

- Incorporates lessons learned from exercises and real-world catastrophes.

- Articulates how we should ensure our long-term success by strengthening the homeland security foundation we have built.

National Strategy for Homeland Security Goals

- **Goal 1:** Prevent and disrupt terrorist attacks.

- **Goal 2:** Protect the American people and our critical infrastructure and key resource.

- **Goal 3:** Respond to and recover from incidents that do occur.

- **Goal 4:** Continue to strengthen the foundation to ensure our long-term success.

The National Response Framework supports Goal #3

Response Doctrine: Overview

Response doctrine defines basic roles, responsibilities, and operational concepts for response across all levels of government and with the private sector and nongovernmental organizations.

It is important to remember that the overarching objective of response activities is life safety, followed by protecting property and the environment.

Response Doctrine: Engaged Partnership

Engaged partnership means that leaders at all levels develop shared response goals and align capabilities so that no one is overwhelmed in times of crisis.

Effective response activities begin with a host of preparedness activities conducted well in advance of an incident. Preparedness involves a combination of planning, resources, training, exercising, and organizing to build, sustain, and improve operational capabilities.

Key Concept: Engaged partnerships are essential to preparedness.

Engaged Partnership: Best Practice #1

For many people, pets are part of the family. Every day, companion animals provide vital services. In the case of farming, animals play a valuable role in the agricultural economy of the community. Studies show that up to 60 percent of pet owners may not evacuate unless they can take their pet along.

In many jurisdictions, such as the city of Lansing, Michigan, Animals in Disaster Planning Taskforces are working to train local workers and volunteers to set up and operate emergency pet shelters in a disaster. The Taskforce is also developing plans to assist pet owners who were not able to evacuate their pets before the disaster.

The Animals in Disaster Planning Taskforce in Lansing is a true partnership and includes the local and county emergency management agencies, Capital Area Humane Society, Ingham County Animal Control, Michigan State University Veterinary Clinic, and Mid-Michigan Red Cross.

Engaged Partnership: Best Practice #2

In the aftermath of the 2007 firestorms in Southern California, the Business Executives for National Security (BENS) served as a crucial bridge between the public and private sectors.

Private-sector liaisons from BENS's Bay Area Business Force and Los Angeles Business Force/Homeland Security Advisory Council worked side-by-side with Federal, State, and local officials inside emergency operations centers to quickly match local needs with business assets.

In that role, Business Force staff helped route millions of dollars worth of food and supplies to affected areas. Just as importantly, they helped prevent major duplications in requests for and delivery of emergency donations and volunteers.

Response Doctrine: Tiered Response

Incidents begin and end locally, and most are managed at the local level.

Many incidents require unified response from local agencies, the private sector, and nongovernmental organizations. Other incidents may require additional support from neighboring jurisdictions or the State.

A small number require Federal support. National response protocols recognize this and are structured to provide additional, tiered levels of support.

Key Concept: A basic premise of the Framework is that incidents are generally handled at the lowest jurisdictional level possible.

Tiered Response: Best Practice

Mutual aid is provided through the Mutual Aid Box Alarm System (MABAS). More than 11,000 of the State of Illinois's 12,000 fire agencies belong to MABAS. There are also MABAS-affiliated agencies in Wisconsin, Indiana, and Missouri. Throughout the Great Lakes region, the groundwork is being laid to establish a compatible mutual aid system modeled after MABAS.

Since the late 1960s, MABAS has provided needed resources to non-declared incidents such as extra-alarm fires, multiple-victim accidents, technical rescues, and hazmat accidents.

MABAS may be activated for larger incidents under the Illinois Emergency Management Agency's statewide mutual aid plan. The statewide plan allows for resource deployment to a stricken area while leaving at least 80 percent of local resources in place to respond to ongoing, routine local emergencies.

Response Doctrine: Scalable, Flexible, and Adaptable Operational Capabilities

The number, type, and sources of resources must be able to expand rapidly to meet needs associated with a given incident.

The Framework builds on the National Incident Management System (NIMS). Together, the Framework and NIMS help to ensure that all response partners use standard command and management structures that allow for scalable, flexible, and adaptable operational capabilities.

Scalable, Flexible, and Adaptable Operational Capabilities: Best Practice

Based on lessons learned from the 9/11 attacks, the New York City Fire Department (FDNY) established and trained Incident Management Teams (IMTs).

The IMTs are designed to provide operational capabilities to ensure that the department has adequate around-the-clock coverage during prolonged incidents. Each team member is trained and credentialed to assume specific Incident Command System (ICS) functions.

When Hurricane Katrina made landfall in August 2005, FDNY had enough trained personnel to deploy a full IMT to assist in the response.

Key Concept: As incidents change in size, scope, and complexity, the response must adapt to meet requirements.

Response Doctrine: Unity of Effort Through Unified Command

Success requires unity of effort, which respects the chain of command of each participating organization while harnessing seamless coordination across jurisdictions in support of common objectives.

As a team effort, unified command allows all agencies with jurisdictional authority and/or functional responsibility for the incident to provide joint support through mutually developed incident objectives and strategies. Each participating agency maintains its own authority, responsibility, and accountability.

Unified Command: Additional Information

Unified command is an Incident Command System (ICS) application used when more than one agency has jurisdiction or when incidents cross political jurisdictions. Agencies work together through the designated members of the unified command to establish a common set of objectives and strategies and a single Incident Action Plan.

As a team effort, unified command allows all agencies with jurisdictional authority and/or functional responsibility for the incident to provide joint support through mutually developed incident objectives and strategies established at the command level. Each participating agency maintains its own authority, responsibility, and accountability.

Unified Command Benefits

- Collective, strategic approach
- Joint priorities and resource allocation
- Single plan and set of objectives

- Improved information flow and coordination

Unity of Effort Through Unified Command: Best Practice #1

In Boston, planned events can attract crowds of 1 million or more participants and spectators, providing an optimal environment to test and improve disaster plans.

Medical planners, led by Boston Emergency Medical Services, began treating special events as "planned disasters" during preparations for events such as the Boston Marathon.

Agencies and organizations involved adopted the Incident Command System, conducted planning and operations using unified command, and integrated aspects of the region's disaster plans into the event's operations plan.

Unity of Effort Through Unified Command: Best Practice #2

In the aftermath of the devastating winds and flooding from Hurricane Katrina, more than 8.1 million gallons of oil escaped from numerous damaged oil infrastructure sources.

Each responding company had its own incident management teams and incident command posts. An area unified command was established to set the priorities for the incident and ensure that competing demands were resolved for the benefit of the entire response effort. The unified command organizations included industry, State, and Federal representatives.

This system enabled government and industry to execute an effective response and avoid catastrophic pollution levels.

Response Doctrine: Readiness To Act

A forward-leaning posture is imperative for incidents that have the potential to expand rapidly in size, scope, or complexity, and for no-notice incidents.

Once response activities have begun, on-scene actions are based on NIMS principles. An effective national response relies on disciplined processes, procedures, and systems.

Key Concept: Readiness is a collective responsibility. Effective national response depends on our readiness to act.

Readiness To Act: Best Practice #1

Through a public-private partnership, the Linn County, Iowa, Emergency Management Agency and the local nuclear power plant have developed a plan for evacuating individuals with special needs (including the elderly, those with mobility limitations, those on medical assistance devices, etc.) during emergencies or disasters.

The voluntary program compiles information on individuals who feel they may need special assistance and enters this information into a database, where it is crosswalked with the county's Geographic Information System (GIS). This allows emergency personnel to quickly determine the location and specific needs of individuals during a disaster.

Readiness To Act: Best Practice #2

In Washington State, the King County Office of Emergency Management, in collaboration with other regional emergency management agencies, a local radio station, the Seattle Mariners baseball team, nongovernmental organizations, and the private sector, promoted community preparedness through the 3 Days, 3 Ways, Are You Ready? campaign.

The program combined the use of print advertisements, outreach activities, and giveaways.

Implementing the Response Doctrine

The response doctrine is not just a philosophy, but rather actions we all can take to be better prepared. To support the doctrine, the National Response Framework is a compendium of resources, not just a single document.

National Response Framework Organization

The National Response Framework is comprised of the core document, the Emergency Support Function (ESF), Support, and Incident Annexes, and the Partner Guides. The core document describes the doctrine that guides our national response, roles and responsibilities, response actions, response organizations, and planning requirements to achieve an effective national response to any incident that occurs.

The following documents provide more detailed information to assist practitioners in implementing the Framework:

- **Emergency Support Function Annexes** group Federal resources and capabilities into functional areas that are most frequently needed in a national response (e.g., Transportation, Firefighting, Search and Rescue).

- **Support Annexes** describe essential supporting aspects that are common to all incidents (e.g., Financial Management, Volunteer and Donations Management, Private-Sector Coordination).

- **Incident Annexes** address the unique aspects of how we respond to seven broad incident categories (e.g., Biological, Nuclear/Radiological, Cyber, Mass Evacuation).

- **Partner Guides** provide ready references describing key roles and actions for local, tribal, State, Federal, and private-sector response partners.

Lesson: **2 - Roles and Responsibilities**

Lesson Overview

This lesson provides an overview of the roles and responsibilities of key partners at the local, tribal, State, and Federal levels who implement the Framework. This includes an important role for the private sector and nongovernmental organizations. At the end of this lesson, you will be able to describe:

- The roles and responsibilities of response partners.

- The process for requesting assistance.

Response Partnerships: Scenario

Effective response requires partnerships among levels of government, the private sector, and nongovernmental organizations. The following scenario illustrates how the various response partners work together within the National Response Framework.

The earthquake comes without warning as residents go about their business on a typical weekday. Measuring 7.5 on the Richter scale, the quake occurs along a fault in the Bay Area. Ninety minutes later, an 8.0 magnitude aftershock occurs along a second fault line to the east.

The effects of the two shocks are profound—10 times more powerful than the 1989 Loma Prieta quake. More than 100,000 people are injured and thousands more are missing. Because the quakes occur during the day, most families in the area are separated.

Damage to bridges as well as public transportation systems essentially severs access to the hardest hit areas. More than 46,000 buildings, including 37 percent of homes, have been destroyed.

The magnitude of this incident results in a nationwide response. As soon as they are able, neighbors rush to help neighbors, and office workers guide their colleagues to safety from unstable buildings.

Under the direction of Incident Commanders, first responders begin critical lifesaving operations, while the local Emergency Operations Center, or EOC, provides support by assessing damage reports and obtaining needed resources. The Mayor's Policy Group is stood up. As representatives of the city's response agencies, this group ensures coordination of response efforts across departments.

The Mayor asks the Governor to declare a state of emergency. At first report of the earthquake, the Governor and emergency management staff began activating State response plans. While the Mayor leads the local response, the Governor has overall responsibility for the public safety and welfare of the residents and provides needed resources and capabilities.

Despite preparedness for earthquakes, the scale of this disaster clearly exceeds the capability of local responders and the State government. The Governor activates existing mutual aid agreements with other States. The Governor also requests a Presidential disaster declaration.

The Secretary of Homeland Security, as the principal Federal official for domestic incident management, immediately begins to coordinate supporting Federal operations. As the State begins to identify the range of its requirements for support, Federal departments and agencies activate their emergency response plans, surge their operations centers, and coordinate their activities through the regional and national coordination centers. FEMA deploys an Incident Management Assistance Team to the State EOC and begins establishing the field structures that will ultimately become the Joint Field Office and coordinate Federal response efforts in the field.

All aspects of the Federal response are organized using the principle of Unified Command, thus allowing various Federal departments and agencies to support State and local responders in a coordinated and unified manner.

The President responds to the Governor's request and issues a major disaster declaration under the Robert T. Stafford Disaster Relief Act. In the declaration, the President designates a Federal Coordinating Officer and provides Federal resources and funds to help support the response and recovery.

The private sector plays many important roles during the response. Activating their emergency plans, businesses begin contacting employees to ensure their well being. Next, the private sector assesses the damage and begins working to restore essential community services, with priority given to restoring infrastructure and providing key resources.

Voluntary and nongovernmental organizations activate their response plans and mobilize their networks to provide requested assistance through the emergency management structures. By integrating into the response structures, these groups provide invaluable capabilities and resources to support the response.

Although just a scenario, this disaster unfortunately is all too plausible. The National Response Framework clearly defines the roles of response partners so that we can effectively work together to meet these challenges.

Partnerships

The video presented a scenario illustrating how response partners work together to meet incident management challenges.

The next information reviews the response roles and responsibilities.

Local Governments

Local jurisdictions are responsible for ensuring the public safety and welfare of their residents.

Local police, fire, emergency medical services, public health and medical providers, emergency management, public works, environmental response professionals, and others in the community are often the first to detect a threat or hazard, or respond to an incident. They also are often the last to leave an incident site or otherwise to cope with the effects of an incident.

Local Governments: Key Players

Local governments manage the vast majority of incidents that incidents occur each day. Local key players include:

Chief Elected or Appointed Official

Role: A mayor, city manager, or county manager, as a jurisdiction's chief executive officer, is responsible for ensuring the public safety and welfare of the people of that jurisdiction.

Responsibilities:

- Establish strong working relationships with local jurisdictional leaders and core private-sector organizations, voluntary agencies, and community partners. The objective is to establish relationships, coordinate, and train with local partners in advance of an incident and to develop mutual aid and/or assistance agreements for support in response to an incident.

- Lead and encourage local leaders to focus on preparedness by participating in planning, training, and exercises.

- Support participation in local mitigation efforts within the jurisdiction including, as appropriate, the private sector.

- Understand and implement laws and regulations that support emergency management and response.

- Ensure that local emergency plans take into account the needs of:

  The jurisdiction, including persons, property, and structures.

    o Individuals with special needs, including those with service animals.

    o Individuals with household pets.

    o Encourage residents to participate in volunteer organizations and training courses.

Emergency Manager

Role: The local emergency manager has the day-to-day authority and responsibility for overseeing emergency management programs and activities.

Responsibilities:

- Coordinate the emergency planning process and work cooperatively with other local agencies and private-sector and nongovernmental organizations.

- Develop mutual aid and assistance agreements.

- Develop and execute public awareness and education programs.

- Conduct exercises to test plans and systems and incorporate lessons learned into the jurisdiction's emergency plan.

- Involve the private sector and nongovernmental organizations in planning, training, and exercises.

- Coordinate damage assessments during an incident.

- Advise and inform local officials about emergency management activities during an incident.

Department and Agency Heads

Role: Department and agency heads collaborate with the emergency manager during development of local emergency plans and provide key response resources.

Responsibilities:

- Participate in the planning process to build specific capabilities (e.g., firefighting, law enforcement, emergency medical services, public works, environmental and natural resources agencies).

- Integrate capabilities into a workable plan to safeguard the community.

- Develop internal policies and procedures to meet response and recovery needs safely.

- Train personnel and participating in interagency training and exercises.

- When an incident occurs, respond according to emergency plans.

Tribal Governments

Tribal governments are responsible for the public safety and welfare of the people of that tribe. Tribal governments:

- Respond to the same range of emergencies and disasters that other jurisdictions face.

- May request and provide assistance from neighboring jurisdictions under mutual aid and assistance agreements.

Although tribal governments can elect to deal directly with the Federal Government, a State Governor must request a Presidential declaration on behalf of a tribe under the Stafford Act.

Additional information can be found in the Tribal Relations Support Annex.

Nongovernmental Organizations

Nongovernmental and voluntary organizations are essential partners in responding to incidents. Working through emergency operations centers and other structures, nongovernmental and voluntary organizations assist in providing:

- Sheltering, emergency food supplies, counseling services, and other vital services to support response and promote the recovery of disaster victims.

- Specialized services that help individuals with special needs, including those with disabilities.

To engage these key partners most effectively, all levels of governments coordinate with voluntary agencies, existing Voluntary Organizations Active in Disaster (VOADs), community and faith-based organizations, and other entities.

Additional information can be found in the Volunteer and Donations Support Annex.

Individuals and Households

Individuals and households play an important role in the overall emergency management strategy. Community members can contribute by:

- Reducing hazards in and around their homes.

- Preparing an emergency supply kit and household emergency plan.

- Monitoring emergency communications carefully.

- Volunteering with an established organization.

- Enrolling in emergency response training courses.

The Private Sector

Forming the foundation for the health of the Nation's economy, the private sector is a key partner in incident management activities at all levels. The private sector:

- Is responsible for most of the critical infrastructure and key resources in the Nation and thus may require assistance in the wake of a disaster or emergency.

- Provides goods and services critical to the response and recovery process, either on a paid basis or through donations.

Additional information about the role of the private sector can be found in the following documents:

- Private-Sector Coordination Support Annex

- Critical Infrastructure and Key Resources Support Annex

State Governments

During response, States play a key role coordinating resources and capabilities throughout the State and obtaining resources and capabilities from other States.

States have significant resources of their own, including emergency management and homeland security agencies, State police, health agencies, transportation agencies, incident management teams, specialized teams, and the National Guard.

The role of the State government in response is to supplement local efforts before, during, and after incidents.

State Response: Key Players

States support their local governments who are closest to those impacted by incidents. State government key players include:

Governor

Role: Public safety and welfare of a State's citizens are fundamental responsibilities of every Governor. For the purposes of the Framework, any reference to a State Governor also references the chief executive of a U.S. territory.

Responsibilities:

- Coordinate State resources and provide the strategic guidance needed to prevent, mitigate, prepare for, respond to, and recover from incidents of all types.

- In accordance with State law, may be able to make, amend, or suspend certain orders or regulations associated with response.

- Communicate to the public and help people, businesses, and organizations cope with the consequences of any type of incident.

- Command the State military forces (National Guard personnel not in Federal service and State militias).

- Coordinate assistance from other States through interstate mutual aid and assistance compacts, such as the Emergency Management Assistance Compact.

- Request Federal assistance including, if appropriate, a Stafford Act Presidential declaration of an emergency or major disaster, when it becomes clear that State capabilities will be insufficient or have been exceeded.

- Coordinate with impacted tribal governments within the State and initiate requests for a Stafford Act Presidential declaration of an emergency or major disaster on behalf of an impacted tribe when appropriate.

State Homeland Security Advisor

Role: The State Homeland Security Advisor serves as counsel to the Governor on homeland security issues and may serve as a liaison between the Governor's office, the State homeland security structure, the Department of Homeland Security (DHS), and other organizations both inside and outside of the State.

Responsibilities:

- Chair a committee comprised of representatives of relevant State agencies.

- Develop prevention, protection, response, and recovery strategies. This also includes preparedness activities associated with these strategies.

Director, State Emergency Management Agency

Role: All States have laws mandating establishment of a State emergency management agency and the emergency plans coordinated by that agency. The Director of the State emergency management agency ensures that the State is prepared to deal with large-scale emergencies.

Responsibilities:

- Coordinate the State response in any incident.

- Support local governments as needed or requested and coordinate assistance with other States and/or the Federal Government.

Other State Department and Agency Heads

Role: Department and agency heads collaborate with and support the State Emergency Management Director.

Responsibilities:

- Develop, plan, and train to internal policies and procedures to meet response and recovery needs safely.

- Participate in interagency training and exercises to develop and maintain the necessary capabilities.

Federal Government

The Federal Government maintains a wide array of capabilities and resources that can assist State governments in responding to incidents.

When an incident occurs that exceeds or is anticipated to exceed State, tribal, or local resources, the Federal Government may provide resources and capabilities to support the State response.

For incidents involving primary Federal jurisdiction or authorities (e.g., on a military base or a Federal facility or lands), Federal departments or agencies may be the first responders and first line of defense, coordinating activities with State, territorial, tribal, and local partners. The Federal Government also maintains working relationships with the private sector and nongovernmental organizations.

Federal Response: Key Players

The Federal Government's response structures are scalable and flexible – adaptable specifically to the nature and scope of a given incident. Click on each key player to learn more.

National Leadership: The President

- Leads the Federal Government response effort.

- Ensures that the necessary coordinating structures, leadership, and resources are applied quickly and efficiently to large-scale and catastrophic incidents.

- Sets policy for large-scale incidents after consulting with the Homeland Security Council and National Security Council.

Incident Management: Secretary of Homeland Security

- Serves as the principal Federal official for domestic incident management.

- Coordinates the Federal resources utilized in the prevention of, preparation for, response to, or recovery from terrorist attacks, major disasters, or other emergencies.

- Provides the President with an overall architecture for domestic incident management and coordinates the Federal response, when required, while relying upon the support of other Federal partners.

- Contributes elements of the response consistent with the Department of Homeland Security (DHS)'s mission, capabilities, and authorities.

  Note: Federal assistance for incidents that do not require DHS coordination may be led by other Federal departments and agencies consistent with their authorities. The Secretary of Homeland Security may monitor such incidents and may activate specific Framework mechanisms to provide support to departments and agencies without assuming overall leadership for the Federal response to the incident.

Incident Management: FEMA Administrator

- Serves as the principal advisor to the President, the Secretary of Homeland Security, and the Homeland Security Council on all matters regarding emergency management.

- Assists the Secretary of Homeland Security to prepare for, protect against, respond to, and recover from all-hazards incidents.

- Manages the operation of the National Response Coordination Center and provides for the effective support of all Emergency Support Functions.

- Makes recommendations to the President through the Secretary of Homeland on Stafford Act declaration requests.

- Manages the core DHS grant programs supporting homeland security.

Law Enforcement: Attorney General

- Serves as the chief law enforcement officer of the United States.

- Generally acting through the Federal Bureau of Investigation:

- Assumes lead responsibility for criminal investigations of terrorist acts or terrorist threats by individuals or groups inside the United States or directed at U.S. citizens or institutions abroad.

- Coordinates activities of the other members of the law enforcement community to detect, prevent, and disrupt terrorist attacks against the United States.

- Approves requests submitted by State Governors pursuant to the Emergency Federal Law Enforcement Assistance Act for personnel and other Federal law enforcement support during incidents.

- Enforces Federal civil rights laws and provides expertise to ensure that these laws are appropriately addressed.

- National Defense and Defense Support of Civil Authorities: Secretary of Defense

- Approves requests for response resources.

The primary mission of the Department of Defense (DOD) and its components is national defense. Because of this critical role, resources are committed after approval by the Secretary of Defense or at the direction of the President. Many DOD components and agencies are authorized to respond to save lives, protect property and the environment, and mitigate human suffering under imminently serious conditions, as well as to provide support under their separate established authorities, as appropriate. The provision of defense support is evaluated by its legality, lethality, risk, cost, appropriateness, and impact on readiness.

- Retains command of military forces.

  When Federal military and civilian personnel and resources are authorized to support civil authorities, command of those forces will remain with the Secretary of Defense. DOD elements in the incident area of operations and National Guard forces under the command of a Governor will coordinate closely with response organizations at all levels.

International Coordination: Secretary of State

- Manages international preparedness, response, and recovery activities relating to domestic incidents.

- Manages efforts related to the protection of U.S. citizens and U.S. interests overseas.

Intelligence: Director of National Intelligence

- Leads the Intelligence Community and serves as the President's principal intelligence advisor.

- Oversees and directs the implementation of the National Intelligence Program.

Other Federal Department and Agency Heads

- Serve in primary, coordinating, and/or support roles based on their authorities and resources and the nature of the threat or incident.

- Participate as members of the Unified Coordination Group in situations where their agency or department has responsibility for directing or managing a major aspect of a response.

- Execute their own authorities to declare disasters or emergencies. For example, the Secretary of Health and Human Services can declare a public health emergency. These declarations may be made independently or as part of a coordinated Federal response. Where those declarations are part of an incident requiring a coordinated Federal response, those Federal departments or agencies act within the overall coordination structure of the Framework.

Note: When the overall coordination of Federal response activities is required, it is implemented through the Secretary of Homeland Security consistent with Homeland Security Presidential Directive (HSPD) 5. Other Federal departments and agencies carry out their response authorities and responsibilities within this overarching construct. Nothing in the Framework alters or impedes the ability of Federal, State, tribal, or local departments and agencies to carry out their specific authorities or perform their responsibilities under all applicable laws, Executive orders, and directives. Additionally, nothing in the Framework is intended to impact or impede the ability of any Federal department or agency to take an issue of concern directly to the President or any member of the President's staff.

Mutual Aid and Assistance Agreements

According to NIMS, mutual aid and assistance agreements should include:

- Definitions of key terms used in the agreement.

- Roles and responsibilities of individual parties.

- Procedures for requesting and providing assistance.

- Procedures, authorities, and rules for allocation and reimbursement of costs.

- Notification procedures.

- Protocols for interoperable communications.

- Relationships with other agreements among jurisdictions.

- Treatment of workers' compensation, liability, and immunity.

- Recognition of qualifications and certifications.

State Assistance to Local Governments

States provide the majority of the external assistance to local jurisdictions. The State is the gateway to several government programs that help communities prepare.

When an incident grows beyond the capability of a local jurisdiction, and responders cannot meet the needs with mutual aid and assistance resources, the local emergency manager contacts the State.

Immediate State Response Activities

Upon receiving a request for assistance from a local government, immediate State response activities may include:

- Coordinating warnings and public information through the activation of the State's public communications strategy and the establishment of a Joint Information Center.

- Distributing supplies stockpiled to meet the emergency.

- Providing needed technical assistance and support to meet the response and recovery needs of individuals and households.

- The Governor suspending existing statutes, rules, ordinances, and orders for the duration of the emergency, to the extent permitted by law, to ensure timely performance of response functions.

- Implementing State donations management plans and coordinating with nongovernmental organizations and the private sector.

- Ordering the evacuation of persons from any portions of the State threatened by the incident, giving consideration to the requirements of special needs populations and those with household pets or service animals.

- Mobilizing resources to meet the requirements of people with special needs, in accordance with the State's preexisting plan and in compliance with Federal civil rights laws.

In addition to these actions, the Governor may activate elements of the National Guard. The National Guard is a crucial State resource, with expertise in communications, logistics, search and rescue, and decontamination. National Guard forces employed under State Active Duty or Title 32 status are under the command and control of the Governor of their State and are not part of Federal military response efforts.

Title 32 Full-Time National Guard Duty refers to Federal training or other duty, other than inactive duty, performed by a member of the National Guard. Title 32 is not subject to posse *comitatus* restrictions and allows the Governor, with the approval of the President or the Secretary of Defense, to order a Guard member to duty to:

- Perform training and other operational activities.

- Conduct homeland defense activities for the military protection of the territory or domestic population of the United States, or of the infrastructure or other assets of the United States determined by the Secretary of Defense to be critical to national security, from a threat or aggression against the United States.

In rare circumstances, the President can federalize National Guard forces for domestic duties under Title 10 (e.g., in cases of invasion by a foreign nation, rebellion against the authority of the United States, or where the President is unable to execute the laws of the United States with regular forces (10 U.S.C. 12406)). When mobilized under Title 10 of the U.S. Code, the forces are no longer under the command of the Governor. Instead, the Department of Defense assumes full responsibility for all aspects of the deployment, including command and control over National Guard forces.

State-to-State Assistance

If additional resources are required, the State often requests assistance from other States by using interstate mutual aid and assistance agreements such as the Emergency Management Assistance Compact (EMAC).

Administered by the National Emergency Management Association, EMAC is a congressionally ratified organization that provides form and structure to the interstate mutual aid and assistance process. Through EMAC or other mutual aid or assistance agreements, a State can request and receive assistance from other member States.

Federal Assistance: Overview

Federal support to States and local jurisdictions takes many forms. The most widely known authority under which assistance is provided for major incidents is the Stafford Act.

In this section, you'll learn how the National Response Framework applies to both Stafford Act and non-Stafford Act incidents, including when one Federal department or agency is called on to support another.

Federal Assistance: Overview Video Transcript

The Federal Government maintains a wide array of capabilities and resources. During this presentation we'll review the various mechanisms within the National Response Framework for providing Federal support.

Perhaps the most widely known authority under which Federal assistance is provided for major incidents is the Stafford Act. In fact, Federal disaster assistance is often thought of as synonymous with Presidential declarations and the Stafford Act. However, Federal assistance under the Stafford Act is only available when the incident exceeds State, tribal, and local resources.

In those circumstances, a Governor may ask the President to declare an emergency or major disaster. Before making a declaration request, the Governor must activate the State's emergency plan and ensure that all appropriate State and local actions have been taken or initiated. Examples of these actions include surveying the affected areas to determine the extent of private and public damage, and conducting joint Preliminary Damage Assessments with FEMA officials to estimate the types and extent of Federal disaster assistance required.

The Governor's request is made through the FEMA Regional Administrator and includes information on the extent and nature of State resources that have been or will be used; a certification by the Governor that State and local governments will assume all applicable non-Federal costs required by the Stafford Act; an estimate of the types and amounts of supplementary Federal assistance required; and designation of the State Coordinating Officer.

The FEMA Regional Administrator evaluates the damage and requirements for Federal assistance and makes a recommendation to the FEMA Administrator. The FEMA Administrator, acting through the Secretary of Homeland Security, then recommends a course of action to the President. In extraordinary circumstances, the President may unilaterally make such a declaration to expedite the delivery of lifesaving assistance.

Following a Presidential declaration, the President appoints a Federal Coordinating Officer to execute Stafford Act authorities. The Federal Coordinating Officer represents the FEMA Administrator in the field and uses the structures and process specified in the National Response Framework to manage the response and recovery efforts.

While the Stafford Act may be the most familiar mechanism for Federal support, it is not the only one.

Often, Federal assistance does not require coordination by the Department of Homeland Security and can be provided without a Presidential emergency or major disaster declaration. In these instances, Federal departments and agencies provide assistance to States, as well as directly to tribes and local jurisdictions, consistent with their own authorities.

It is important to note that the National Response Framework does not alter or impede the ability of Federal departments and agencies to carry out their specific response authorities. For example, local and tribal governments can request assistance directly from the Environmental Protection Agency and the U.S. Coast Guard under the Comprehensive Environmental Response, Compensation, and Liability Act.

In addition, Federal departments and agencies routinely manage the response to incidents under their statutory or executive authorities. An example of such an authority is the wildland firefighting support provided by the U.S. Forest Service. When assistance is being coordinated by a Federal agency with primary jurisdiction, the Department of Homeland Security may activate Framework mechanisms to support the response without assuming overall leadership for the incident.

In addition to providing assistance to local, tribal, and State governments, the National Response Framework allows for Federal-to-Federal support. A Federal entity with primary responsibility and authority for handling an incident may request Federal assistance under the National Response Framework. In these circumstances, the Department of Homeland Security coordinates the response using multiagency coordination structures established in the Framework.

The National Response Framework provides structures for implementing nationwide response policy and operational coordination for all types of domestic incidents. Given its flexibility and scalability, the Framework can be implemented to deliver the resources at the needed level of coordination across a wide range of authorities.

Requesting Federal Assistance: Summary

The Framework is intended to strengthen, organize, and coordinate response actions across all levels. The doctrine of tiered response emphasizes that response to incidents should be handled at the lowest jurisdictional level capable of handling the work.

The following resources provide additional information on requesting assistance:

- Overview of Stafford Act Support to States

- Overview of Federal-to-Federal Support

- Financial Management Support Annex

Lesson: **3 - Response Actions**

Lesson Overview

The Framework is intended to strengthen, organize, and coordinate response actions across all levels. This lesson describes and outlines key tasks related to the three phases of effective response: prepare, respond, and recover.

At the end of this lesson, you will be able to describe the actions that support national response.

Response Actions: Video Transcript

Responders and emergency managers are both doers and planners, which means to lead response and recovery efforts effectively, they must also prepare effectively. In this segment, we'll look at how the National Response Framework strengthens our ability to prepare for, respond to, and recover from incidents.

Let's begin with how we prepare as a Nation. Preparedness is essential for effective response. During preparedness, response partners plan; organize, train, and equip; exercise; and evaluate and improve.

Planning includes the development of policies, plans, procedures, mutual aid and assistance agreements, strategies, and other arrangements to perform missions and tasks. Plans should address all hazards and be tailored to each jurisdiction.

Organizing to execute response activities includes developing an overall organizational structure, strengthening leadership at each level, and assembling well-qualified teams of paid and volunteer staff for essential response and recovery tasks. The National Incident Management System, or NIMS, provides standard command and management structures used during response. These common structures enable responders from different jurisdictions and disciplines to work together to respond to incidents.

Once responders are equipped with resources, training helps build essential response capabilities and readiness.

Exercises provide opportunities to test plans and improve proficiency. When response partners exercise together, interagency coordination and communications are improved and capability gaps and opportunities for improvement are identified.

Evaluation and continuous improvement are cornerstones of effective preparedness. Corrective action programs help response partners to evaluate response operations, capture lessons learned, and make improvements.

Once an incident occurs, priorities shift – from building capabilities to taking actions to save lives, protect property and the environment, and preserve the social, economic, and political structure of the jurisdiction. Let's look at the four key actions that typically occur in support of a response.

Situational awareness requires continuous monitoring of relevant sources of information regarding actual and developing incidents. For an effective national response, jurisdictions must continuously refine their ability to assess the situation as an incident unfolds and rapidly provide accurate and accessible information to decisionmakers. It is essential that all response partners develop a common operating picture and synchronize their response operations and resources.

When an incident occurs, responders assess the situation, identify and prioritize requirements, and activate available resources and capabilities to save lives, protect property and the environment, and meet basic human needs.

Response is guided by the common principles, structures, and coordinating processes established by NIMS.

Demobilization is the orderly, safe, and efficient return of a resource to its original location and status. Demobilization begins as soon as possible to facilitate accountability of the resources and to make resources available for other incidents as needed.

Once immediate lifesaving activities are complete, the focus shifts to recovery. During short-term recovery, basic services and functions are restored. In the long term, recovery is a restoration of both the personal lives of individuals and the livelihood of the community.

Each member of our society, including our leaders, professional emergency managers, private-sector representatives, and nongovernmental organizations plays a vital role in strengthening the Nation's response capabilities. The partnerships fostered by the National Response Framework greatly improve our ability to work together to prepare, respond, and recover.

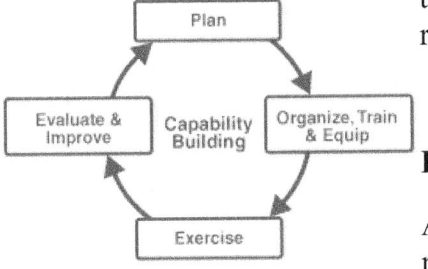

Preparedness Cycle

As you learned in the video presentation, preparedness is essential for effective response.

This section briefly reviews the six essential activities for preparing to respond to an incident: plan, organize, train, equip, exercise, and evaluate and improve.

Plan

Planning makes it possible to manage the entire life cycle of a potential crisis, determine capability requirements, and help response partners learn their roles. In addition, planning:

- Includes the collection and analysis of intelligence and information, as well as the development of policies, plans, procedures, mutual aid and assistance agreements, strategies, and other arrangements to perform missions and tasks.

- Improves effectiveness by clearly defining required capabilities, shortening the time required to gain control of an incident, and facilitating the rapid exchange of information.

Make sure that your plans:

- Are developed using hazard identification and risk assessment methodologies.

- Are all-hazards in scope while providing hazard-specific guidance.

- Define leadership roles and responsibilities.

- Identify the decisions that need to be made, who will make them, and when.

- Are integrated, operational, and incorporate key private-sector and nongovernmental elements.

- Include provisions for all persons, including special needs populations and those with household pets.

- Address all contingencies across various scenarios, including no-notice and forewarned events.

- Are augmented by specific procedures and protocols to guide rapid implementation.

- Are living documents that are updated continuously based on changing conditions and lessons learned.

Organize, Train, and Equip

Organize: Executing response activities includes developing an overall organizational structure, strengthening leadership at each level, and assembling well-qualified teams of paid and volunteer staff for essential response and recovery tasks.

Train: Building essential response capabilities nationwide requires a systematic program to train individual teams and organizations – to include governmental, nongovernmental, private-sector, and voluntary organizations – to meet a common baseline of performance and certification standards.

Equip: It is critical to acquire equipment that will perform to established standards, including the capability to be interoperable with equipment used by other jurisdictions and/or participating organizations.

National Incident Management System

The National Incident Management System (NIMS) provides a comprehensive framework to ensure that responders from across the country are organized, trained, and equipped in a manner that allows them to work together seamlessly. Response organizations should:

- Conduct a thorough, systematic inventory of their response resources.

- Establish resource management practices that conform to NIMS.

- Have a cadre of personnel (which can include full-time employees, temporary or reserve personnel, and contractors) who are trained in incident management and response principles and organized into teams.

- Pre-position resources close to those localities most at risk.

Organize, Train, and Equip: Best Practice

Partnering with FEMA and the National Voluntary Organizations Active in Disaster (National VOAD), the Points of Light Foundation & Hands-On Network assists in the management of unaffiliated volunteers in disasters.

This network concluded that establishing a volunteer reception center within an emergency plan is essential to managing the large numbers of people who volunteer their services after a major incident. Although the public is encouraged to volunteer with a local agency before an emergency, unaffiliated volunteers will inevitably offer to help.

A volunteer reception center allows volunteers to be organized, credentialed, and tracked throughout their time of service.

Exercise

Exercises provide opportunities to test plans and improve proficiency in a risk-free environment. Effective exercises:

- Assess and validate proficiency levels.

- Clarify and familiarize personnel with roles and responsibilities.

- Improve interagency coordination and communications, highlight capability gaps, and identify opportunities for improvement.

Exercise Requirements

Local, tribal, State, and Federal jurisdictions should exercise their own response capabilities and evaluate their abilities to perform expected responsibilities and tasks. This is a basic responsibility of all entities and is distinct from participation in other interagency exercise programs.

In addition, the Department of Homeland Security (DHS) coordinates the National Exercise Program, which requires Federal departments and agencies to participate in an exercise program based upon the National Planning Scenarios contained in the National Preparedness Guidelines. This program coordinates and, where appropriate, integrates a 5-year homeland security exercise schedule across Federal agencies and incorporates exercises at the State and local levels.

Criteria for Effective Exercises

Exercises should:

- Include multidisciplinary, multijurisdictional incidents.

- Include participation of private-sector and nongovernmental organizations.

- Cover aspects of preparedness plans, particularly the processes and procedures for activating local, intrastate, or interstate mutual aid and assistance agreements.

- Contain a mechanism for incorporating corrective actions.

Exercise: Best Practice #1

The New Jersey Business Force (NJBF) is a private, nonprofit enterprise dedicated to emergency preparedness and homeland security through the development of public-private partnerships.

One initiative being worked through this partnership is the rapid distribution of medical supplies to hospitals and the general public during a public health emergency. Tabletop exercises are being used to:

- Explore how private-sector resources can be integrated into the New Jersey Department of Health's medical supply distribution plan.

- Develop a workable plan for private-sector involvement in mass statewide medical supply distribution.

Exercise: Best Practice #2

The Arizona Division of Emergency Management (ADEM) conducts annual emergency response exercises to test incident preparedness and response measures. With more than 700 individuals and 70 organizations participating, these annual exercises strengthen partnerships among State and local agencies and nongovernmental organizations.

ADEM exercise planners incorporated Community Emergency Response Teams (CERTs) into recent exercises. During exercises, CERT volunteers applied their training by following Incident Command System practices and following orders correctly.

The exercises demonstrated the potential benefits of these partnerships. By effectively using CERT volunteers, responders found that they were able to concentrate on those tasks for which they are uniquely trained, such as fighting fires, providing emergency medical care, or securing dangerous areas.

Evaluate and Improve

Evaluation and continual process improvement are cornerstones of effective preparedness. Upon concluding an exercise, jurisdictions should:

- Evaluate performance against relevant capability objectives and identify gaps.

- Develop corrective action plans with specific recommendations for changes in practice, timelines for implementation, and assignments for completion.

All local, tribal, State, and Federal entities should institute a corrective action program to evaluate exercise participation and response, capture lessons learned, and make improvements in their response capabilities.

Homeland Security Exercise and Evaluation Program

The Homeland Security Exercise and Evaluation Program (HSEEP):

- Is a capabilities- and performance-based exercise program.

- Establishes standardized policy, methodology, and language for designing, developing, conducting, and evaluating all exercises.

- Provides tools and resources including policy and guidance, training, technology, and direct exercise support.

The HSEEP Toolkit is a Web-based system that enables implementation of the corrective action program process.

Response

Depending on the size, scope, and magnitude of an incident, communities, States, and, in some cases, the Federal Government will respond.

Gain and Maintain Situational Awareness

Situational awareness requires continuous monitoring of relevant sources of information regarding actual and developing incidents.

The scope and type of monitoring vary based on the type of incidents being evaluated and needed reporting thresholds. Critical information is passed through established reporting channels according to established security protocols.

Situational Awareness Priorities

When developing protocols that promote situational awareness, priority should be given to:

- Providing the right information at the right time.

- Improving and integrating national reporting.

- Linking operations centers and tapping subject-matter experts.

- Standardizing reporting.

Gain and Maintain Situational Awareness: Best Practice

The Mid-America Regional Council (MARC) serves as the association of city and county governments and the metropolitan planning organization for the bi-State Kansas City metropolitan region. MARC has developed a system that allows emergency responders to share information and to coordinate planning and responses for large-scale incidents by:

- Connecting approximately 100 government agencies and private organizations across 8 counties and 2 States in the Kansas City region.

- Allowing agencies throughout the region to share data on emergency personnel, plans, and assets.

- Sharing purchasing guidelines and fund management information.

Activate and Deploy Resources and Capabilities

At the onset of an incident or planned event, responders:

- Assess the situation.

- Identify and prioritize requirements.

- Establish incident objectives.

- Activate available resources and capabilities to save lives, protect property and the environment, and meet basic human needs.

At the scene, Incident Commanders develop Incident Action Plans while local, tribal, State, and/or Federal Government entities develop plans for coordinating support efforts.

Initial Response Actions

Examples of initial response actions include:

- **Activating people, resources, and other capabilities.** Initial actions include the systematic activation of people and teams and establishment of incident management and response structures to organize and coordinate an effective response. The resources and capabilities deployed and the activation of supporting incident management structures should be directly related to the size, scope, nature, and complexity of the incident. All responders should maintain and regularly exercise notification systems and protocols.

- **Requesting additional resources and capabilities.** Additional resources may be requested through mutual aid and assistance agreements, the State, or the Federal Government. For all incidents, especially large-scale incidents, it is essential to prioritize and clearly communicate incident requirements so that resources can be efficiently matched, typed, and mobilized to support operations.

- **Identifying needs and pre-positioning resources.** When planning for heightened threats or in anticipation of large-scale incidents, local and tribal jurisdictions, States, and the Federal Government should identify needed resources and capabilities. Based on asset availability, resources should be pre-positioned and response teams and other support resources may be placed on alert or deployed to a staging area.

Mobilization and deployment will be most effective when supported by planning that includes:

- **Pre-Scripted Mission Assignments.** The Federal Government and many State governments use pre-scripted mission assignments to assist in planning and to reduce the time it takes to deploy response resources. Pre-scripted mission assignments identify resources or capabilities of government organizations that are commonly called upon during response to an incident.

- **Advanced Readiness Contracting.** Advanced readiness contracting ensures that contracts are in place before an incident for commonly needed commodities and services such as ice, water, plastic sheeting, temporary power, and debris removal. This type of contracting improves the ability to secure supplies and services by streamlining the process of ordering, acquiring, and distributing resources when needed.

Coordinate Response Actions

Coordination of response activities:

- Is enhanced by the use NIMS common principles, structures, and coordinating processes.

- Involves the clear delegation of assigned roles and responsibilities.

- Requires that critical information is provided through established reporting mechanisms.

Sample Response Actions:

- Warning the public and providing accessible emergency public information.

- Implementing evacuation and sheltering plans that include provisions for special needs populations and household pets.

- Sheltering evacuees in preidentified, physically accessible shelters.

- Providing food, water, and other necessities to meet the needs of all people, including persons with disabilities and other special needs.

- Performing search and rescue.

- Treating the injured.

- Providing law enforcement and investigation.

- Controlling hazards (extinguishing fires, containing hazardous materials spills, etc.).

- Ensuring responder safety and health.

Managing Response Actions

Response actions are managed using the NIMS Command and Management component. This component includes the following elements: the Incident Command System, Multiagency Coordination Systems, and Public Information.

NIMS Command and Management Elements

Incident Command System

The Incident Command System (ICS) is a standardized, on-scene, all-hazard incident management approach that:

- Allows for the integration of facilities, equipment, personnel, procedures, and communications operating within a common organizational structure.

- Enables a coordinated response among various jurisdictions and functional agencies, both public and private.

- Establishes common processes for planning and managing resources.

Multiagency Coordination Systems

The second Command and Management element is Multiagency Coordination Systems. Multiagency coordination is a process that allows all levels of government and all disciplines to work together more efficiently and effectively.

A multiagency coordination system is not simply a physical location or facility. Rather, a multiagency coordination system is a process that:

- Defines business practices, standard operating procedures, processes, and protocols by which participating agencies will coordinate their interactions.

- Supports the Incident Command/Unified Command (who manage the incident scene) by providing coordination and assistance with making policy-level decisions.

Public Information

The final Command and Management element is Public Information.

Public information consists of the processes, procedures, and systems to communicate timely, accurate, and accessible information on the incident's cause, size, and current situation to the public, responders, and additional stakeholders (both directly affected and indirectly affected).

Public information must be coordinated and integrated across jurisdictions and across agencies/organizations; among Federal, State, tribal, and local governments; and with the private sector and nongovernmental organizations.

Public information, education strategies, and communications plans help ensure that numerous audiences receive timely, consistent messages about:

- Lifesaving measures.
- Evacuation routes.
- Threat and alert system notices.
- Other public safety information.

Demobilize

Incident managers should plan and prepare for the demobilization process at the same time that they begin the resource mobilization process.

Early planning for demobilization facilitates accountability and makes the logistical management of resources as efficient as possible – in terms of both costs and time of delivery.

Recover

After immediate lifesaving activities are complete, the focus shifts to assisting individuals, households, critical infrastructure, and businesses in meeting basic needs and returning to self-sufficiency.

Short-term recovery is immediate and overlaps with response. Recovery actions include providing essential public health and safety services, restoring interrupted utility and other essential services, reestablishing transportation routes, and providing food and shelter for those displaced by the incident. Although called "short term," some of these activities may last for weeks.

Long-term recovery, which is outside the scope of the National Response Framework, may involve some of the same actions but may continue for a number of months or years.

Support Annexes

Numerous procedures and administrative functions are required to support incident management. The National Response Framework includes Support Annexes that describe overarching actions applicable to nearly every type of incident.

- **Critical Infrastructure and Key Resources Support Annex**
  The Critical Infrastructure and Key Resources Support Annex describes policies, roles and responsibilities, and the concept of operations for assessing, prioritizing, protecting, and restoring critical infrastructure and key resources (CIKR) of the United States and its territories and possessions during actual or potential domestic incidents. The annex details processes to ensure coordination and integration of CIKR-related activities among a wide array of public and private incident managers and CIKR security partners within immediate incident areas as well as at the regional and national levels.

- **Financial Management Support Annex**
  The Financial Management Support Annex provides basic financial management guidance for all participants in NRF activities. This includes guidance for all Federal departments and agencies providing assistance for incidents requiring a coordinated Federal response. The financial management

function is a component of Emergency Support Function #5 – Emergency Management. The processes and procedures described in this annex ensure that funds are provided expeditiously and that financial operations are conducted in accordance with established Federal law, policies, regulations, and standards.

- **International Coordination Support Annex**
  The International Coordination Support Annex provides guidance on carrying out responsibilities for international coordination in support of the Federal Government's response to a domestic incident with an international component.

- **Private-Sector Coordination Support Annex**
  The Private-Sector Coordination Support Annex describes the policies, responsibilities, and concept of operations for Federal incident management activities involving the private sector during incidents requiring a coordinated Federal response. In this context, the annex further describes the activities necessary to ensure effective coordination and integration with the private sector, both for-profit and not-for-profit, including the Nation's critical infrastructure, key resources, other business and industry components, and not-for-profit organizations, including those serving special needs populations, engaged in response and recovery.

- **Public Affairs Support Annex**
  The Public Affairs Support Annex describes the interagency policies and procedures used to rapidly mobilize Federal assets to prepare and deliver coordinated and sustained messages to the public in response to incidents requiring a coordinated Federal response.

- **Tribal Relations Support Annex**
  The Tribal Relations Support Annex describes the policies, responsibilities, and concept of operations for effective coordination and interaction of Federal incident management activities with those of tribal governments and communities during incidents requiring a coordinated Federal response. The processes and functions described in this annex help facilitate

the delivery of incident management programs, resources, and support to tribal governments and individuals.

- **Volunteer and Donations Management Support Annex**
  The Volunteer and Donations Management Support Annex describes the coordination processes used to support the State in ensuring the most efficient and effective use of unaffiliated volunteers, unaffiliated organizations, and unsolicited donated goods to support all Emergency Support Functions for incidents requiring a Federal response, including offers of unaffiliated volunteer services and unsolicited donations to the Federal Government.

- **Worker Safety and Health Support Annex**
  The Worker Safety and Health Support Annex provides Federal support to Federal, State, tribal, and local response and recovery worker safety and health during incidents requiring a coordinated Federal response. The annex, coordinated by the Department of Labor/Occupational Safety and Health Administration (OSHA), describes the technical assistance resources, capabilities, and other support to ensure that response and recovery worker safety and health risks are anticipated, recognized, evaluated, communicated, and consistently controlled.

Lesson: 4 - Response Organization

Lesson Overview

This lesson explains how we as a Nation are organized to implement response actions. At the end of this lesson, you will be able to identify:

- The organizational structures that have been developed, tested, and refined over time and how these structures are applied at all levels to support an effective response.

- The key staff positions needed to operate this system and their relationships and dependencies.

Response Organization Overview: Video Transcript

The National Response Framework integrates organizational structures that have been developed, tested, and refined over time. In this presentation, we'll review the major response organizations used at all levels, beginning with local response organizations.

A basic premise of the Framework is that incidents are generally handled at the lowest jurisdictional level possible. Incidents begin and end locally. And most incidents are managed entirely at the local level.

Local responders use the Incident Command System, or ICS, to manage response operations. ICS is a management system designed to enable effective incident management by integrating a combination of facilities, equipment, personnel, procedures, and communications operating within a common organizational structure.

The Incident Commander communicates with the local emergency operations center, or EOC, to report on the incident status and request resources. During an incident, the local emergency manager ensures the EOC is staffed to support the incident command and arranges needed resources. The chief elected or appointed official provides policy direction and supports the Incident Commander and emergency manager, as needed.

When an incident grows beyond the capability of a local jurisdiction, and responders cannot meet the needs with mutual aid and assistance resources, the local emergency manager may contact the State. State

EOCs are activated as necessary to ensure that responders have the resources they need. The Governor may provide the needed resources or request assistance from other States through mutual aid and assistance agreements such as the Emergency Management Assistance Compact.

When it is clear that State capabilities will be exceeded, the Governor may request Federal assistance. Federal assistance can be provided to State, tribal, and local jurisdictions, and to other Federal agencies, in a number of different ways through various mechanisms and authorities.

For our purposes, let's assume the Governor is requesting assistance under the Stafford Act. In this case, the FEMA Regional Administrator deploys a liaison to the State EOC to provide technical assistance and fully activates the Regional Response Coordination Center, or RRCC. The RRCC coordinates Federal regional response efforts until the Joint Field Office is established. The Joint Field Office, or JFO, is the primary Federal incident management field structure. The JFO provides a central location for the coordination of Federal, State, tribal, and local governments and private-sector and nongovernmental organizations with primary responsibility for response and recovery.

Although the JFO uses an ICS structure, it does not manage on-scene operations. Rather, the JFO provides support to on-scene efforts. The JFO is led by the Unified Coordination Group, which is typically comprised of the Federal Coordinating Officer, who is appointed by the President to execute Stafford Act authorities; the State Coordinating Officer, who is appointed by the Governor to coordinate State disaster assistance efforts; and others, such as the Senior Health Official, Department of Defense representative, or Senior Federal Law Enforcement Official.

For a catastrophic or complex incident, a Principal Federal Official, or PFO, may be appointed to serve as the Secretary of Homeland Security's representative. When appointed, the PFO works within the Unified Coordination Group and interfaces with all levels of responders regarding the overall Federal incident management strategy but does not direct nor replace the incident command structure established at the incident.

At the national level, the President ensures the necessary coordinating structure, leadership, and Federal resources are directed quickly and efficiently to large-scale incidents. The Secretary of Homeland Security is the principal Federal official responsible for domestic incident management. The FEMA Administrator serves as the principal advisor to the President and the Secretary of Homeland Security and is responsible for the preparation for, protection against, response to, and recovery from all-hazards incidents.

To ensure integration of Federal response efforts, the National Operations Center, called the NOC, serves as the primary national hub for situational awareness and operations coordination. The NOC provides the Secretary of Homeland Security and other principals with information necessary to make critical national-level incident management decisions.

One key component of the NOC is the National Response Coordination Center, which is referred to as the NRCC. The NRCC is FEMA's focal point for national resource coordination. The NRCC provides overall emergency management coordination, conducts operational planning, deploys national-level teams, and builds and maintains a common operating picture.

This presentation introduced the major response organizations at the local, regional, field, and national levels. By promoting the use of these flexible, scalable, and adaptable structures, the National Response Framework ensures that we are prepared to respond, together as a Nation.

National Mandate

Homeland Security Presidential Directive (HSPD) 5 called for a single, comprehensive system to enhance the ability of the United States to manage domestic incidents.

As presented earlier, the National Incident Management System (NIMS) provides a consistent nationwide template to enable all levels of government, the private sector, and nongovernmental organizations to work together during an incident.

In the remainder of this lesson, we will review the organizational structures outlined in the video presentation.

Local Response Structure

Local responders use the Incident Command System (ICS) to establish standardized organizational structures.

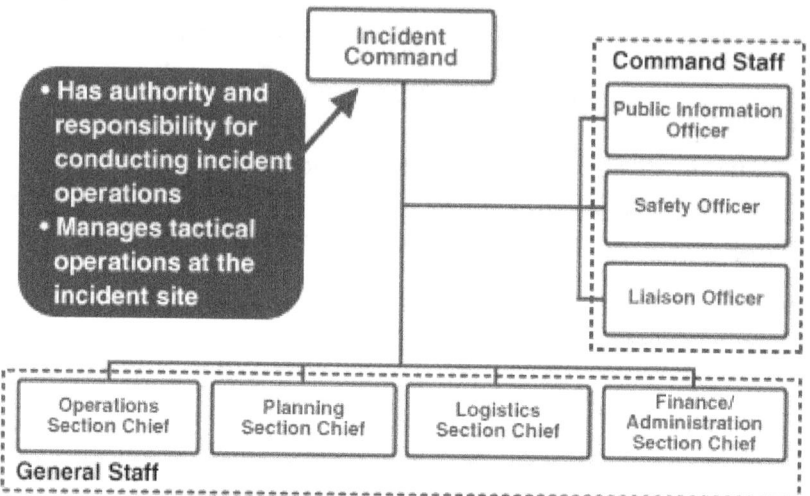

Incident Commander

The Incident Commander is the individual responsible for all incident activities, including the development of strategies and tactics and the ordering and the release of resources. The Incident Commander has overall authority and responsibility for conducting incident operations and is responsible for the management of tactical operations at the incident site.

The Incident Commander must:

- Have clear authority and know agency policy.

- Ensure incident safety.

- Establish the Incident Command Post.

- Set priorities, and determine incident objectives and strategies to be followed.

- Establish the Incident Command System organization needed to manage the incident.

- Approve the Incident Action Plan.

- Coordinate Command and General Staff activities.

- Approve resource requests and use of volunteers and auxiliary personnel.

- Order demobilization as needed.

- Ensure after-action reports are completed.

- Authorize information released to the media.

Command Staff

In an Incident Command organization, the Command Staff typically includes the following personnel:

- The **Public Information Officer** is responsible for interfacing with the public and media and/or with other agencies with incident-related information requirements.

- The **Safety Officer** monitors incident operations and advises the Incident Commander/Unified Command on all matters relating to operational safety, including the health and safety of emergency responder personnel.

- The **Liaison Officer** is the point of contact for representatives of other governmental agencies, nongovernmental organizations, and/or private entities.

Additional Command Staff positions may be added depending upon incident needs and requirements

General Section Staff

- **Operations Section Chief:** The Operations Section Chief is responsible to the Incident Commander/Unified Command for the direct management of all incident-related operational activities. The Operations Section Chief will establish tactics for the assigned operational period and be directly involved in development of the Incident Action Plan. An Operations Section Chief should be designated for each operational period.

- **Planning Section Chief:** The Planning Section Chief oversees the collection, evaluation, and dissemination of the incident situation information and intelligence for the Incident Commander/Unified Command and incident management personnel. The Planning Section then prepares status reports, displays situation information, maintains the status of resources assigned to the incident, and prepares and documents the Incident Action Plan, based on Operations Section input and guidance from the Incident Commander/Unified Command.

- **Logistics Section Chief:** The Logistics Section Chief is responsible for all service support requirements needed to facilitate effective and efficient incident management, including ordering resources from off-incident locations. The Logistics Section also provides facilities, security (of the Incident Command facilities), transportation, supplies, equipment maintenance and fuel, food services, communications and information technology support, and emergency responder medical services, including inoculations, as required.

- **Finance/Administration Section Chief**: A Finance/Administration Section Chief is assigned when the incident management activities require on-scene or incident-specific finance and other administrative support services. Some of the functions that fall within the scope of this Section are recording personnel time, maintaining vendor contracts, overseeing compensation and claims, and conducting an overall cost analysis for the incident. When a Finance/Administration Section is established, close coordination with the Planning Section and Logistics Section is also essential so that operational records can be reconciled with financial documents. In addition to monitoring multiple sources of funds, the Section Chief must track and report to the Incident Command the accrued cost as the incident progresses. This allows the Incident Command to forecast the need for additional funds before operations are affected negatively.

Unified Command

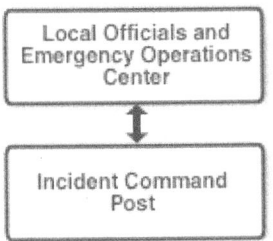

As an incident expands in complexity, a Unified Command may be established. In a Unified Command, individuals designated by their jurisdictional or organizational authorities (or by departments within a single jurisdiction) work together to:

- Determine objectives, strategies, plans, resource allocations, and priorities.

- Develops a single Incident Action Plan.

- Execute integrated incident operations and maximize the use of assigned resources.

Area Command

Area Command is an organization that oversees the management of multiple incidents that are each being handled by a separate command organization.

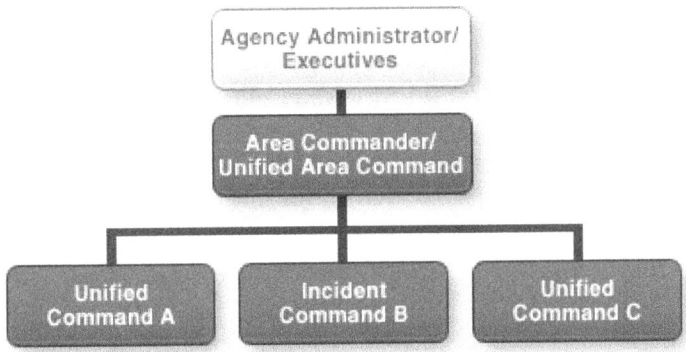

An Area Command is activated only if necessary, depending on the complexity of the incident and incident management span-of-control considerations.

Incident Command Post

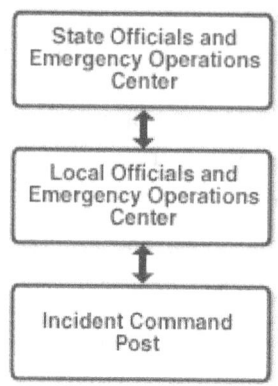

State Officials and Emergency Operations Center

Local Officials and Emergency Operations Center

Incident Command Post

The Incident Command and management organization is located at the Incident Command Post. The Incident Command directs operations from the Incident Command Post, which is generally located at or in the immediate vicinity of the incident site.

As emergency management/response personnel deploy, they must, regardless of agency affiliation, report to and check in at the designated location and receive an assignment in accordance with the established procedures.

Local Emergency Operations Center

An emergency operations center (EOC) is the physical location at which the coordination of information and resources to support incident management (on-scene operations) activities normally takes place. During an incident, the local EOC supports the on-scene response by relieving the burden of external coordination and securing additional resources.

EOC core functions include coordination; communications; resource allocation and tracking; and information collection, analysis, and dissemination.

EOCs may be staffed by personnel representing multiple jurisdictions, the private sector, and nongovernmental organizations.

State Emergency Operations Center

State emergency operations centers (EOCs) are activated as necessary to support local EOCs. The key function of State EOC personnel is to ensure that those who are located at the scene have the resources (e.g., personnel, tools, and equipment) they need for the response.

The State EOC is the central location from which off-scene activities supported by the State are coordinated. Chief elected and appointed officials are located at the State EOC, as well as personnel supporting core functions.

Joint Information Center

In order to coordinate the release of emergency information and other public affairs functions, a Joint Information Center (JIC) may be established.

The JIC serves as a focal point for coordinated and timely release of incident-related information to the public and the media. Information about where to receive assistance is communicated directly to victims and their families in an accessible format and in appropriate languages for those with limited English proficiency.

JICs may be established at all levels of government, at incident sites, or can be components of Multiagency Coordination Systems (i.e., emergency operations centers and the Joint Field Office).

Federal Response Structures

The next section presents the following levels of Federal response structures:

- National Policy Level
- National and Regional Coordination
- Field Operations

National Response Policy

As you learned earlier, the President leads the Nation in responding to large-scale incidents. The following entities provide policy-level guidance on national response issues:

- **Homeland Security Council and National Security Council:** The Homeland Security Council (HSC) and National Security Council (NSC) advise the President on national strategic issues and policy during large-scale incidents. The HSC and NSC ensure coordination for all homeland and national security-related activities among executive departments and agencies and promote effective development and implementation of related policy. The HSC and NSC ensure unified leadership across the Federal Government.

- **Domestic Readiness Group:** The Domestic Readiness Group (DRG) is an interagency body convened on a regular basis to develop and coordinate preparedness, response, and incident management policy. This group evaluates various policy issues of interagency importance regarding domestic preparedness and incident management and makes recommendations to senior levels of the policymaking structure for decision. During an incident, the DRG may be convened by DHS to evaluate relevant interagency policy issues regarding response and develop recommendations as may be required.

- **Counterterrorism Security Group:** The Counterterrorism Security Group (CSG) is an interagency body convened on a regular basis to develop terrorism prevention policy and to coordinate threat response and law enforcement investigations associated with terrorism. This group evaluates various policy issues of interagency importance regarding counterterrorism and makes recommendations to senior levels of the policymaking structure for decision.

Domestic Incident Management

The Department of Homeland Security (DHS) coordinates Federal incident management activities in support of our State and local partners. The following DHS key players have significant roles in coordinating incident management:

- **Secretary of Homeland Security:** The Secretary serves as the principal Federal official for domestic incident management, which includes coordinating both Federal operations within the United States and Federal resources used in response to or recovery from terrorist attacks, major disasters, or other emergencies. The Secretary of Homeland Security is by Presidential directive and statutory authority also responsible for coordination of Federal resources utilized in the prevention of, preparation for, response to, or recovery from terrorist attacks, major disasters, or other emergencies, excluding law enforcement responsibilities otherwise reserved to the Attorney General.

- **FEMA Administrator:** The FEMA Administrator is the principal advisor to the President, the Secretary of Homeland Security, and the Homeland Security Council regarding emergency management. The FEMA Administrator's duties include operation of the National Response Coordination Center, the effective support of all Emergency Support Functions, and, more generally, preparation for, protection against, response to, and recovery from all-hazards incidents. Reporting to the Secretary of Homeland Security, the Administrator also is responsible for management of the core DHS grant programs supporting homeland security.

- **DHS Director of Operations Coordination:** The DHS Director of Operations Coordination is the Secretary's principal advisor for the overall departmental level of integration of incident management operations and oversees the National Operations Center. Run by the Director, the National Operations Center is intended to provide a one-stop information source for incident information sharing with the White House and other Federal departments and agencies at the headquarters level.

National Operations Center

To ensure integration of Federal response efforts, the National Operations Center (NOC) serves as the primary national hub for situational awareness and operations coordination.

The NOC provides the Secretary of Homeland Security and other principals with information necessary to make critical national-level incident management decisions.

National Response Coordination Center

One key component of the National Operations Center is the National Response Coordination Center (NRCC).

The NRCC:

- Is FEMA's focal point for national resource coordination.

- Provides overall emergency management coordination.

- Conducts operational planning.

- Deploys national-level teams.

- Builds and maintains a common operating picture.

National Response Coordination Center: Additional Information

The National Response Coordination Center (NRCC) is FEMA's primary operations management center, as well as the focal point for national resource coordination. As a 24/7 operations center, the NRCC monitors potential or developing incidents and supports the efforts of regional and field components.

The NRCC also has the capacity to increase staffing immediately in anticipation of or in response to an incident by activating the full range of Emergency Support Functions and other personnel as needed to provide resources and policy guidance to a Joint Field Office or other local incident management structures.

The NRCC provides overall emergency management coordination, conducts operational planning, deploys national-level entities, and collects and disseminates incident information as it builds and maintains a common operating picture. Representatives of nonprofit organizations within the private sector may participate in the NRCC to enhance information exchange and cooperation between these entities and the Federal Government.

National Infrastructure Coordinating Center

The goal of the National Infrastructure Protection Plan is to build a safer, more secure, and more resilient America by enhancing protection of the Nation's critical infrastructure and key resources.

The National Infrastructure Coordinating Center (NICC), another NOC component, monitors the Nation's critical infrastructure and key resources on an ongoing basis. During an incident, the NICC provides a coordinating forum to share information across infrastructure and key resources sectors through appropriate information-sharing entities.

See the Critical Infrastructure and Key Resources Support Annex and learn more.

Additional Federal Operations Centers

The Federal Government has a wide range of headquarters-level operations centers that maintain situational awareness within their functional areas and provide relevant information to the NOC.

Following are examples of other Federal operations centers:

- **National Military Command Center:** The National Military Command Center (NMCC) is the Nation's focal point for continuous monitoring and coordination of worldwide military operations. It directly supports combatant commanders, the Chairman of the Joint Chiefs of Staff, the Secretary of Defense, and the President in the command of U.S. Armed Forces in peacetime contingencies and war. Structured to support the President and Secretary of Defense effectively and efficiently, the Center participates in a wide variety of activities, ranging from missile warning and attack assessment to management of peacetime contingencies such as Defense Support of Civil Authorities (DSCA) activities. In conjunction with monitoring the current worldwide situation, the Center alerts the Joint Staff and other national agencies to developing crises and will initially coordinate any military response required.

- **National Counterterrorism Center:** The National Counterterrorism Center (NCTC) serves as the primary Federal organization for integrating and analyzing all intelligence pertaining to terrorism and counterterrorism and for conducting strategic operational planning by integrating all instruments of national power.

- **Strategic Information and Operations Center:** The FBI Strategic Information and Operations Center (SIOC) is the focal point and operational control center for all Federal intelligence, law enforcement, and investigative law enforcement activities related to domestic terrorist incidents or credible threats, including leading attribution investigations. The SIOC serves as an information clearinghouse to help collect, process, vet, and disseminate information relevant to law enforcement and criminal investigation efforts in a timely manner. The SIOC maintains direct connectivity with the NOC. The SIOC, located at FBI Headquarters, supports the FBI's mission in leading efforts of the law enforcement community to detect, prevent, preempt, and disrupt terrorist attacks against the United States.

The SIOC maintains liaison with the National Joint Terrorism Task Force (NJTTF). The mission of the NJTTF is to enhance communications, coordination, and cooperation among Federal, State, tribal, and local agencies representing the intelligence, law enforcement, defense, diplomatic, public safety, and homeland security communities by providing a point of fusion for terrorism intelligence and by supporting Joint Terrorism Task Forces throughout the United States.

- **Other DHS Operations Centers:** Depending upon the type of incident (e.g., National Special Security Events), the operations centers of other DHS operating Components may serve as the primary operations management center in support of the Secretary. These are the U.S. Coast Guard, Transportation Security Administration, U.S. Secret Service, and U.S. Customs and Border Protection operations centers.

Regional Response Coordination Center

Each of FEMA's regional offices maintains a Regional Response Coordination Center (RRCC). The RRCCs are coordination centers that expand to become an interagency facility in anticipation of a serious incident or immediately following an incident.

Operating under the direction of the FEMA Regional Administrator, the RRCCs coordinate Federal regional response efforts until the Joint Field Office is established.

FEMA Regional Offices

FEMA has 10 regional offices, each headed by a Regional

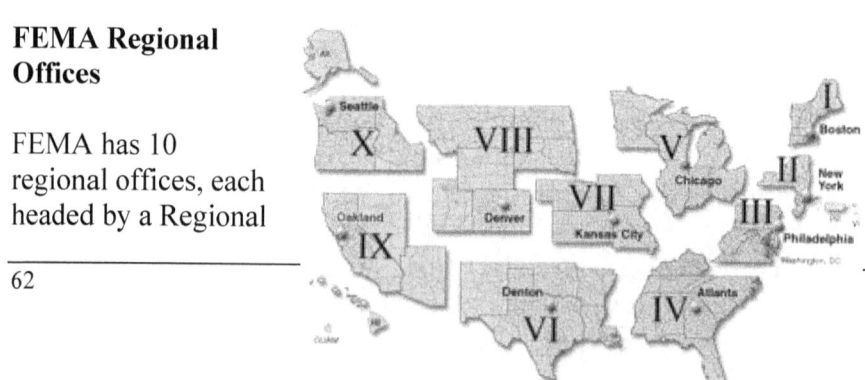

Administrator. The regional field structures are FEMA's permanent presence for communities and States across the Nation. The staff at these offices support development of all-hazards operational plans and generally help States and communities become better prepared. These regional offices mobilize Federal assets and evaluation teams to work with State and local agencies. Many of FEMA's most experienced response personnel are employed at regional offices.

Each of FEMA's regional offices maintains a Regional Response Coordination Center (RRCC). The RRCCs are 24/7 coordination centers that expand to become an interagency facility staffed by Emergency Support Functions in anticipation of a serious incident in the region or immediately following an incident. Operating under the direction of the FEMA Regional Administrator, the RRCCs coordinate Federal regional response efforts, and maintain connectivity with State EOCs, State fusion centers, Federal Executive Boards, and other Federal and State operations and coordination centers that have potential to contribute to development of situational awareness. Ongoing RRCC operations transition to a JFO once it is established, so that the RRCC can remain ready to deal with new incidents.

Other Federal departments and agencies have regional or field offices that may participate with State, tribal, and local governments in planning for incidents under their jurisdiction and provide initial response assets to the incident.

Initial Federal Response Assets

FEMA liaisons coordinate with the State to ensure that needed initial Federal assets are dispatched before or during the first hours of an incident.

Examples of initial Federal response assets include:

- **Incident Management Assistance Teams:** Incident Management Assistance Teams (IMATs) are interagency, regionally based response teams that provide a forward Federal presence to improve response to serious incidents.

  The IMATs support efforts to meet the emergent needs of State and local jurisdictions, possess the capability to provide initial situational awareness for Federal decisionmakers, and support the establishment of Federal coordination efforts with the State.

- **Hurricane Liaison Team:** The Hurricane Liaison Team (HLT) is a small team designed to enhance hurricane disaster response by facilitating information exchange between the National Hurricane Center in Miami and other National Oceanic and Atmospheric Administration components, as well as Federal, State, tribal, and local government officials.

- **Urban Search and Rescue Task Forces:** The National Urban Search and Rescue (US&R) Response System is a framework for structuring local emergency services personnel into integrated response task forces. The 28 National US&R Task Forces, complete with the necessary tools, equipment, skills, and techniques, can be deployed by FEMA to assist State, tribal, and local governments in rescuing victims of structural collapse incidents or to assist in other search and rescue missions. Each task force must have all its personnel and equipment at the embarkation point within 6 hours of activation. The task force can be dispatched and en route to its destination within a matter of hours.

- **Mobile Emergency Response Support:** The primary function of Mobile Emergency Response Support (MERS) is to provide mobile telecommunications capabilities and life, logistics, operational, and power-generation support required for the on-site management of response activities. MERS support falls into three broad categories: (1) operational support elements, (2) communications equipment and operators, and (3) logistics support.

  MERS supports Federal, State, tribal, and local responders in their efforts to save lives, protect property, and coordinate response operations. Staged in six strategic locations, one with offshore capabilities, the MERS detachments can concurrently support multiple field operating sites within an incident area.

Proactive Response to Catastrophic Incidents

Prior to and during catastrophic incidents, especially those that occur with little or no notice, the State and Federal governments may take proactive measures to mobilize and deploy assets in anticipation of a formal request from the State for Federal assistance.

Such deployments of significant Federal assets would likely occur for catastrophic events involving chemical, biological, radiological, nuclear, or high-yield explosive weapons of mass destruction, large-magnitude earthquakes, or other catastrophic incidents affecting heavily populated areas.

Joint Field Office

The Joint Field Office (JFO) is the primary Federal incident management field structure.

The JFO provides a central location for the coordination of Federal, State, tribal, and local governments and private-sector and nongovernmental organizations with primary responsibility for response and recovery.

Although the JFO uses an ICS structure, it does not manage on-scene operations. Rather, the JFO provides support to on-scene efforts.

Unified Coordination Group

The Unified Coordination Group provides leadership within the Joint Field Office. The Unified Coordination Group:

- Is comprised of senior leaders representing State and Federal interests, and in certain circumstances tribal governments, local jurisdictions, the private sector, or nongovernmental organizations.

- Applies unified command principles to coordinating assistance being provided to support the local, tribal, and State response.

Unified Coordination Group Members

The composition of the Unified Coordination Group varies, depending upon the scope and nature of the incident. For a Stafford Act incident, two key group members include:

- **Federal Coordinating Officer (FCO).** The FCO is appointed by the President to execute Stafford Act authorities. The FCO is the primary Federal representative with whom the State, tribal, and local response officials interface to identify needs and set objectives for an effective collaborative response.

- **State Coordinating Officer (SCO).** The SCO is appointed by the Governor to coordinate State disaster assistance efforts. The SCO works with the FCO to formulate State requirements and set priorities for use of Federal support.

Additional Unified Coordination Group Members

Additional members of the Unified Coordination Group may include the following individuals:

- **Federal Resource Coordinator:** In non-Stafford Act situations, when a Federal department or agency acting under its own authority has requested the assistance of the Secretary of Homeland Security to obtain support from other Federal departments and agencies, DHS may designate a Federal Resource Coordinator (FRC). In these situations, the FRC coordinates support through interagency agreements and memorandums of understanding. Relying on the same skill set, DHS may select the FRC from the Federal Coordinating Officer cadre or other personnel with equivalent knowledge, skills, and abilities. The FRC is responsible for coordinating timely delivery of resources to the requesting agency.

- **Senior Federal Law Enforcement Official:** The Senior Federal Law Enforcement Official (SFLEO) is an individual appointed by the Attorney General during an incident requiring a coordinated Federal response to coordinate all law enforcement, public safety, and security operations with intelligence or investigative law enforcement operations directly related to the incident. The SFLEO is a member of the

Unified Coordination Group and, as such, is responsible to ensure that allocation of law enforcement requirements and resource allocations are coordinated as appropriate with all other members of the Group. In the event of a terrorist incident, the SFLEO will normally be a senior FBI official who has coordinating authority over all law enforcement activities related to the incident, both those falling within the Attorney General's explicit authority as recognized in HSPD-5 and those otherwise directly related to the incident itself.

- **Defense Coordinating Officer:** The Department of Defense (DOD) has appointed 10 Defense Coordinating Officers (DCOs) and assigned one to each FEMA region. If requested and approved, the DCO serves as DOD's single point of contact at the JFO for requesting assistance from DOD. With few exceptions, requests for Defense Support of Civil Authorities (DSCA) originating at the JFO are coordinated with and processed through the DCO. The DCO may have a Defense Coordinating Element consisting of a staff and military liaison officers to facilitate coordination and support to activated Emergency Support Functions (ESFs).

  Specific responsibilities of the DCO (subject to modification based on the situation) include processing requirements for military support, forwarding mission assignments to the appropriate military organizations through DOD-designated channels, and assigning military liaisons, as appropriate, to activated ESFs.

  Note: DOD is a full partner in the Federal response to domestic incidents, and its response is fully coordinated through the mechanisms of this Framework. Concepts of "command" and "unity of command" have distinct legal and cultural meanings for military forces and military operations. For Federal military forces, command runs from the President to the Secretary of Defense to the Commander of the combatant command to the DOD on-scene commander. Military forces will always remain under the operational and administrative control of the military chain of command, and these forces are subject to redirection or recall at any time. The ICS "unified command" concept is distinct from the military chain of command use of this term. And, as such, military forces do not operate under the command of the Incident Commander or under the unified command structure.

- **Joint Task Force Commander:** Based on the complexity and type of incident, and the anticipated level of DOD resource involvement, DOD may elect to designate a Joint Task Force (JTF) to command Federal (Title 10) military activities in support of the incident objectives. If a JTF is established, consistent with operational requirements, its command and control element will be co-located with the senior on-scene leadership at the JFO to ensure coordination and unity of effort. The co-location of the JTF command and control element does not replace the requirement for a Defense Coordinating Officer (DCO)/Defense Coordinating Element as part of the JFO Unified Coordination Staff. The DCO remains the DOD single point of contact in the JFO for requesting assistance from DOD.

  The JTF Commander exercises operational control of Federal military personnel and most defense resources in a Federal response. Some DOD entities, such as the U.S. Army Corps of Engineers, may respond under separate established authorities and do not provide support under the operational control of a JTF Commander. Unless federalized, National Guard forces remain under the control of a State Governor. Close coordination between Federal military, other DOD entities, and National Guard forces in a response is critical.

- **Other Senior Officials**: Based on the scope and nature of an incident, senior officials from other Federal departments and agencies; State, tribal, or local governments; and the private sector or nongovernmental organizations may participate in a Unified Coordination Group. Usually, the larger and more complex the incident, the greater the number of entities represented.

Principal Federal Official

For catastrophic or unusually complex incidents, the Secretary of Homeland Security may designate a single Principal Federal Official (PFO) to:

- Serve in the field as his or her primary representative to ensure consistency and effectiveness of Federal support and incident management.

- Interface with Federal, State, tribal, and local jurisdictional officials regarding the overall Federal incident management strategy.

- Provide a primary point of contact and situational awareness locally.

- Act as the primary Federal spokesperson for coordinated media and public communications.

Principal Federal Official and Other Senior Officials

As a member of the Unified Coordination Group, the PFO promotes collaboration and works to resolve any Federal interagency conflict that may arise. The PFO:

- Does not direct nor replace the incident command structure established at the incident.

- Does not have directive authority over the FCO, Senior Federal Law Enforcement Official, DOD Joint Task Force Commander, or any other Federal or State official.

During an incident, the same individual cannot serve as both the PFO and the FCO. When both positions are assigned, the FCO will have responsibility for administering Stafford Act authorities.

Federal Assets Assigned to an Incident Scene

The JFO may assign Federal assets, such as an Urban Search and Rescue Task Force, to assist at an incident scene. In these circumstances, Federal assets are integrated into the unified command structure at the incident scene.

While integrating into tactical operations managed by the on-scene incident command structure, these Federal assets continue to coordinate and communicate critical information to the JFO.

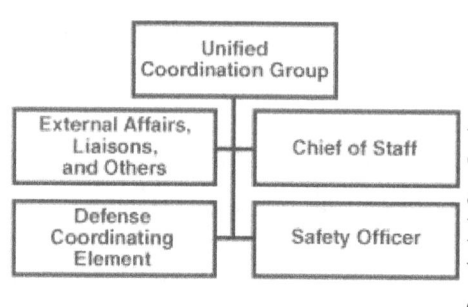

Federal Agencies Responding Under Their Own Authorities

As mentioned earlier in this course, some types of Federal assistance are performed by Federal departments or agencies under their own authorities and do not require Presidential approval.

The jurisdiction or agency with functional or statutory authority manages the incident and establishes the appropriate response structures. Depending on the type and magnitude of the incident, a JFO may or may not be established.

JFO Coordination Staff

Typically, the JFO structure includes a Unified Coordination Staff. The Unified Coordination Group determines the required staffing based on incident needs.

You should refer to the JFO Standard Operating Procedure at the NRF Resource Center for further details on these and other Federal staff positions supporting the field operation.

JFO Sections

The JFO is organized into the following four Sections based on the ICS standard organization:

- Operations Section

- Planning Section

- Logistics Section

- Finance and Administration Section

Remember that although the JFO Sections use ICS titles, their function is to support rather than command the incident.

JFO and Incident Command Sections

Although the JFO Sections use ICS titles, their function is to support rather than command the incident. Below is a summary of the different roles assumed by each Section.

Section	Joint Field Office	Incident Command
Operations	The JFO Operations Section coordinates operational support with on-scene incident management efforts. Branches, Divisions, and Groups may be added or deleted as required, depending on the nature of the incident. The Operations Section is also responsible for coordinating with other Federal facilities that may be established to support incident management activities.	The Incident Command Operations Section is responsible for all tactical incident operations and implementation of the Incident Action Plan. In the Incident Command System, it normally includes subordinate Branches, Divisions, and/or Groups.
Planning	The JFO Planning Section's functions include the collection, evaluation, dissemination, and use of information regarding the threat or incident and the status of Federal resources. The Planning Section prepares and documents Federal support actions and develops unified	The Incident Command Planning Section is responsible for the collection, evaluation, and dissemination of operational information related to the incident, and for the preparation and documentation of the Incident Action Plan. This Section also

	action, contingency, long-term, and other plans.	maintains information on the current and forecasted situation and on the status of resources assigned to the incident.
Logistics	The JFO Logistics Section coordinates logistics support that includes: control of and accountability for Federal supplies and equipment; resource ordering; delivery of equipment, supplies, and services to the JFO and other field locations; facility location, setup, space management, building services, and general facility operations; transportation coordination and fleet management services; information and technology systems services; administrative services such as mail management and reproduction; and customer assistance.	The Incident Command Logistics Section is responsible for providing facilities, services, and material support for the incident. It also provides facilities, security (of the Incident Command facilities), transportation, supplies, equipment maintenance and fuel, food services, communications and information technology support, and emergency responder medical services, including inoculations, as required.
Finance and Administration	The JFO Finance and Administration Section is responsible for the financial management, monitoring, and tracking of all Federal costs relating to the incident and the functioning of the JFO	The Incident Command Finance and Administration Section responsible for all administrative and financial considerations surrounding an incident. Some of the functions

	while adhering to all Federal laws and regulations.	that fall within the scope of this Section are recording personnel time, maintaining vendor contracts, overseeing compensation and claims, and conducting an overall cost analysis for the incident.

Emergency Support Functions: Overview—Video Transcript

Emergency Support Functions, or ESFs, are used by the Federal Government and many States as the primary mechanism to organize and provide assistance.

ESFs are organized into fifteen functional areas such as transportation, public works and engineering, firefighting, search and rescue, mass care, housing, and human services, public health and medical services, agriculture and natural resources, and many more. ESFs may be selectively activated for both Stafford Act and non-Stafford Act incidents and are assigned to support headquarters, regional, and field activities.

At the Joint Field Office, these resources are assigned where needed within the Unified Coordination structure. For example, if a State requests assistance with a mass evacuation, resources from several different ESFs may be integrated into a single branch or group within the Operations Section. Regardless of where ESFs may be assigned, they coordinate closely with one another to accomplish their missions.

National Response Framework Annexes describe the scope, policies, and concept of operations of each ESF. In addition, these annexes identify ESF coordinators, primary agencies, and support agencies. Let's take a closer look at each of these roles.

An ESF coordinator has ongoing management oversight throughout the preparedness, response, and recovery phases of incident management.

A primary agency is a Federal agency with significant authorities, roles, resources, or capabilities for a particular function within an ESF. During a Stafford Act incident, the ESF primary agency serves as a Federal executive agent under the Federal Coordinating Officer.

Support agencies are those entities with specific capabilities or resources that assist the primary agency in executing the mission of the ESF.

Throughout the year, ESFs plan and prepare with all participating organizations and form partnerships with the private sector and nongovernmental organizations. In doing so, Emergency Support Functions are a key element for building our national response capability.

ESF Functional Areas

The ESFs serve as the primary operational-level mechanism to provide assistance in the following functional areas:

ESF #1: Transportation
ESF Coordinator: Department of Transportation

- Aviation/airspace management and control

- Transportation safety

- Restoration and recovery of transportation infrastructure

- Movement restrictions

- Damage and impact assessment

ESF #2: Communications
ESF Coordinator: DHS (National Communications System)

- Coordination with telecommunications and information technology industries

- Restoration and repair of telecommunications infrastructure

- Protection, restoration, and sustainment of national cyber and information technology resources

- Oversight of communications within the Federal incident management and response structures

ESF #3: Public Works and Engineering
ESF Coordinator: Department of Defense (U.S. Army Corps of Engineers)

- Infrastructure protection and emergency repair

- Infrastructure restoration

- Engineering services and construction management

- Emergency contracting support for life-saving and life-sustaining services

ESF #4: Firefighting
ESF Coordinator: Department of Agriculture (U.S. Forest Service)

- Coordination of Federal firefighting activities

- Support to wildland, rural, and urban firefighting operations

ESF #5: Emergency Management
ESF Coordinator: DHS (FEMA)

- Coordination of incident management and response efforts

- Issuance of mission assignments

- Resource and human capital

- Incident action planning

- Financial management

ESF #6: Mass Care, Emergency Assistance, Housing, and Human Services
ESF Coordinator: DHS (FEMA)

- Mass care

- Emergency assistance

- Disaster housing

- Human services

ESF #7: Logistics Management and Resource Support
ESF Coordinator: General Services Administration and DHS (FEMA)

- Comprehensive, national incident logistics planning, management, and sustainment capability

- Resource support (facility space, office equipment and supplies, contracting services, etc.)

ESF #8: Public Health and Medical Services
ESF Coordinator: Department of Health and Human Services

- Public health

- Medical

- Mental health services

- Mass fatality management

ESF #9: Search and Rescue
ESF Coordinator: DHS (FEMA)

- Life-saving assistance

- Search and rescue operations

ESF #10: Oil and Hazardous Materials Response
ESF Coordinator: Environmental Protection Agency

- Oil and hazardous materials (chemical, biological, radiological, etc.) response

- Environmental short- and long-term cleanup

ESF #11: Agriculture and Natural Resources
ESF Coordinator: Department of Agriculture

- Nutrition assistance

- Animal and plant disease and pest response

- Food safety and security

- Natural and cultural resources and historic properties protection

- Safety and well-being of household pets

ESF #12: Energy
ESF Coordinator: Department of Energy

- Energy infrastructure assessment, repair, and restoration

- Energy industry utilities coordination

- Energy forecast

ESF #13: Public Safety and Security
ESF Coordinator: Department of Justice

- Facility and resource security

- Security planning and technical resource assistance

- Public safety and security support

- Support to access, traffic, and crowd control

ESF #14: Long-Term Community Recovery
ESF Coordinator: DHS (FEMA)

- Social and economic community impact assessment

- Long-term community recovery assistance to States, tribes, local governments, and the private sector

- Analysis and review of mitigation program implementation

ESF #15: External Affairs
ESF Coordinator: DHS

- Emergency public information and protective action guidance

- Media and community relations

- Congressional and international affairs

- Tribal and insular affairs

ESF Annexes

The ESF Annexes describe the scope, policies, and concept of operations of each ESF. In addition, these annexes identify:

- An **ESF coordinator**, who has ongoing management oversight throughout the preparedness, response, and recovery phases of incident management.

- A **primary agency**, which is a Federal agency with significant authorities, roles, resources, or capabilities for a particular function within an ESF.

- **Support agencies**, which are those entities with specific capabilities or resources that assist the primary agency in executing the mission of the ESF.

You may access the latest copies of the ESF Annexes at the NRF Resource Center, www.fema.gov/NRF.

ESF Activation

ESFs may be selectively activated for both Stafford Act and non-Stafford Act incidents. Not all incidents requiring Federal support result in the activation of ESFs.

For Stafford Act incidents, the NRCC or RRCC may activate specific ESFs by directing appropriate departments and agencies to initiate the actions delineated in the ESF Annexes.

The ESFs deliver a broad range of technical support and other services at the regional level in the Regional Response Coordination Centers, and in the Joint Field Office and Incident Command Posts, as required by the incident.

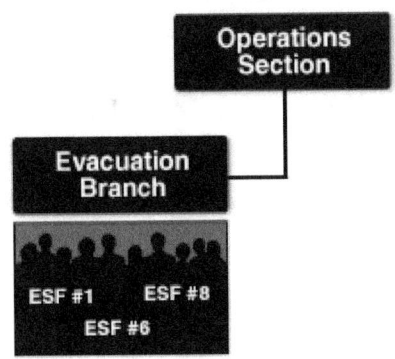

ESFs Within the JFO Structure

Resources coordinated though ESFs are assigned where needed within the response structure.

For example, if a State requests assistance with a mass evacuation, resources from several different ESFs may be integrated into a single Branch or Group within the Operations Section. During the response, these resources would report to a supervisor within the assigned Branch or Group.

Regardless of where ESFs may be assigned, they coordinate closely with one another to accomplish their missions.

Field-Level Structures and Partnerships

The chart below to review the field-level response structures and partnerships.

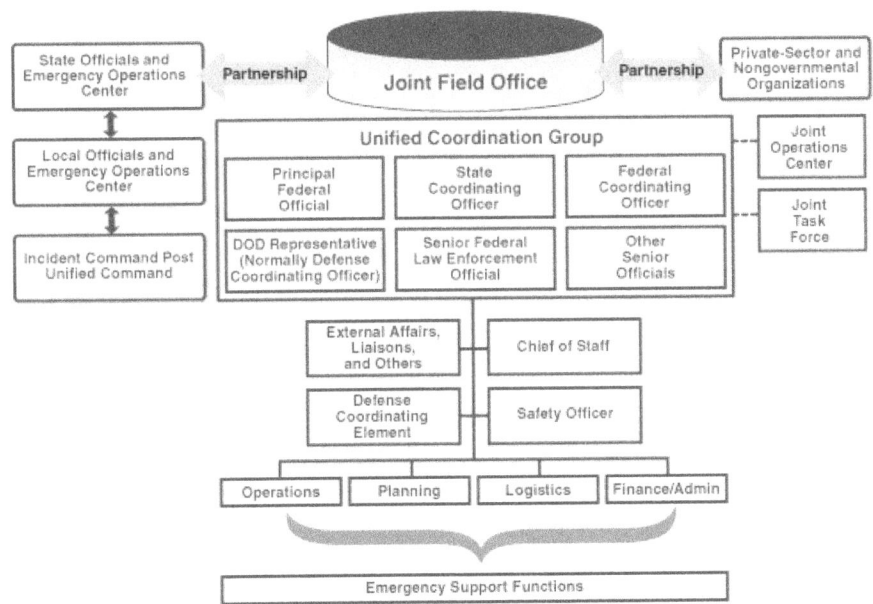

Lesson: **5 - Planning**

Lesson Overview

This lesson summarizes planning structures that are relevant to the National Response Framework. The Framework fosters unity of effort for emergency operations planning by providing common doctrine and purpose.

At the completion of this lesson, you will be able to describe the relationship between planning and national preparedness.

Planning: A Critical Element of Effective Response – Video Transcript

Planning is the cornerstone of national preparedness. The National Response Framework provides a foundation for unified planning for all response partners.

Plans are continuous and evolving. They anticipate actions, maximize opportunities, and guide response operations. That is why plans are best described as "living" documents.

Effective planning allows jurisdictions to influence the course of events by determining actions, policies, and processes in advance of an incident. Planning promotes unity of effort by providing a common blueprint for activity in the event of an emergency.

Emergency planning is a national priority. To address this priority, the National Preparedness Guidelines have been developed.

These Guidelines are comprised of four critical elements. The first element is the National Preparedness Vision, which provides a concise statement of the core preparedness goal for the Nation.

The next element is the National Planning Scenarios, which form a basis for coordinated planning, training, and exercising. These scenarios are planning tools that depict a full range from terrorist attacks to natural disasters.

The third element is the Universal Task List, which provides a menu of unique tasks linked to prevention, protection, response, and recovery strategies. This invaluable resource identifies the critical tasks for which response capabilities must be developed.

The final element is the Target Capabilities List, which defines specific response capabilities that all levels of government should possess.

In addition to these elements, the National Preparedness Guidelines integrate key guidance documents such as: the National Incident Management System, the National Infrastructure Protection Plan, and other national continuity policies and directives.

The Federal planning structure involves the development of three levels of plans for each of the National Planning Scenarios. The first level includes a Strategic Guidance Statement and Strategic Plan. Together these documents define broad national strategic objectives, delineate roles, and establish capabilities and performance measures.

Next is the development of the National-Level Interagency Concept Plan. This plan describes the concept of operations for integrating and synchronizing Federal capabilities.

The third level encompasses Federal department and agency Operations Plans. These plans identify the specific resources, personnel, and assets needed to support the national concept of operations.

The State, tribal, and local planning structure is supported by Federal preparedness assistance.

All levels of government have responsibility to develop detailed, robust, all-hazards plans. These plans are developed using hazard identification and risk assessment methodologies. To ensure that our national planning system is fully integrated, these plans must be tested against all manner and magnitude of threats and hazards.

Planning across the full range of homeland security operations is an inherent responsibility of every level of government. By providing common doctrine and purpose, the National Response Framework lays the foundation for a mutually supportive planning system that fosters engaged partnerships at all levels.

The National Response Framework and Planning

Plans anticipate actions, maximize opportunities, and guide response operations. That is why plans are best described as "living" documents.

Planning across the full range of homeland security operations is an inherent responsibility of every level of government and should include stakeholders from the private sector and nongovernmental organizations.

Next, we'll review the planning structures that are most relevant to the Framework.

National Preparedness Guidelines

Emergency planning is a national priority. To address this priority, the National Preparedness Guidelines have been developed. These Guidelines are comprised of four critical elements:

- The National Preparedness Vision
- The National Planning Scenarios
- The Universal Task List
- The Target Capabilities List

National Preparedness Vision

The National Preparedness Vision provides a concise statement of the core preparedness goal for the Nation. The vision for the National Preparedness Guidelines is:

"A NATION PREPARED with coordinated capabilities to prevent, protect against, respond to, and recover from all hazards in a way that balances risk with resources and need."

National Planning Scenarios

The National Planning Scenarios are planning tools that represent a minimum number of credible scenarios depicting the range of potential terrorist attacks and natural disasters and related impacts facing our Nation. These scenarios form a basis for coordinated Federal planning, training, and exercises.

- **Universal Task List:** The Universal Task List is a menu of unique tasks that link strategies to prevention, protection, response, and recovery tasks for the major events represented by the National Planning Scenarios. The List provides a common vocabulary of critical tasks that support development of essential capabilities among organizations at all levels.

- **Target Capabilities List:** The Target Capabilities List defines specific capabilities that all levels of government should possess in order to respond effectively to incidents.

  Capabilities Definition: The National Preparedness Guidelines define capabilities as providing the means to accomplish a mission or function and achieve desired outcomes by performing critical tasks, under specified conditions, to target levels of performance.

  Each capability includes a description of the major activities performed within the capability and the critical tasks and measures associated with the activity. Critical tasks are those tasks that must be performed during a major event in order to minimize the impact on lives, property, and the economy.

Integrating Other Key Guidance Documents

The National Preparedness Guidelines integrate key guidance documents such as:

- National Incident Management System

- National Infrastructure Protection Plan

- Other National Continuity Policies and Directives

National Planning Scenarios

Let's take a closer look at the National Planning Scenarios. Homeland Security Presidential Directive 8, "National Preparedness," Annex I (National Planning), describes the use of the National Planning Scenarios.

The National Planning Scenarios are the focus of Federal planning efforts. These scenarios represent examples of the gravest dangers facing the United States and have been accorded the highest priority for Federal planning.

Using a shared set of scenarios provides a common yardstick for determining how to achieve expected planning results.

National Planning Scenarios

- Scenario 1: Nuclear Detonation – Improvised Nuclear Device
- Scenario 2: Biological Attack – Aerosol Anthrax
- Scenario 3: Biological Disease Outbreak – Pandemic Influenza
- Scenario 4: Biological Attack – Plague
- Scenario 5: Chemical Attack – Blister Agent
- Scenario 6: Chemical Attack – Toxic Industrial Chemical
- Scenario 7: Chemical Attack – Nerve Agent
- Scenario 8: Chemical Attack – Chlorine Tank Explosion
- Scenario 9: Natural Disaster – Major Earthquake
- Scenario 10: Natural Disaster – Major Hurricane
- Scenario 11: Radiological Attack – Radiological Dispersal Device
- Scenario 12: Explosives Attack – Bombing Using Improvised Explosive Device

- Scenario 13: Biological Attack – Food Contamination

- Scenario 14: Biological Attack – Foreign Animal Disease

- Scenario 15: Cyber Attack

Federal Planning Structure

The Federal planning structure involves the development of the following three levels of plans for each of the National Planning Scenarios:

- **Strategic Guidance Statement and Strategic Plan.** Together these documents define broad national strategic objectives, delineate roles, and establish capabilities and performance measures.

- **National-Level Interagency Concept Plan.** This plan describes the concept of operations for integrating and synchronizing Federal capabilities.

- **Federal Department and Agency Operations Plans.** These plans identify the specific resources, personnel, and assets needed to support the national concept of operations.

State, Tribal, and Local Government Planning

State, tribal, and local governments:

- Have responsibility to develop robust all-hazards plans and hazard- or incident-specific annexes with supporting procedures and protocols to address their locally identified hazards and risks.

- Use hazard identification and risk assessment (HIRA) to identify hazards and associated risks to persons, property, and structures and to improve protection from natural- and human-caused hazards.

In most instances, Federal plans are implemented when a State's resources are not sufficient to cope with an incident and the Governor has requested Federal assistance.

Criteria for Successful Planning

The National Response Framework promotes the use of the following criteria to measure key aspects of response planning:

- **Acceptability:** A plan is acceptable if it can meet the requirements of anticipated scenarios, can be implemented within the costs and timeframes that senior officials and the public can support, and is consistent with applicable laws.

- **Adequacy:** A plan is adequate if it complies with applicable planning guidance, planning assumptions are valid and relevant, and the concept of operations identifies and addresses critical tasks specific to the plan's objectives.

- **Completeness:** A plan is complete if it incorporates major actions, objectives, and tasks to be accomplished. The complete plan addresses the personnel and resources required and sound concepts for how those will be deployed, employed, sustained, and demobilized. It also addresses timelines and criteria for measuring success in achieving objectives, and the desired end state. Completeness of a plan can be greatly enhanced by including in the planning process all those who could be affected.

- **Consistency and Standardization of Products:** Standardized planning processes and products foster consistency, interoperability, and collaboration.

- **Feasibility:** A plan is considered feasible if the critical tasks can be accomplished with the resources available internally or through mutual aid, immediate need for additional resources from other sources (in the case of a local plan, from State or Federal partners) are identified in detail and coordinated in advance, and procedures are in place to integrate and employ resources effectively from all potential providers.

- **Flexibility:** Flexibility and adaptability are promoted by decentralized decisionmaking and by accommodating all hazards ranging from smaller-scale incidents to wider national contingencies.

- **Interoperability and Collaboration:** A plan is interoperable and collaborative if it identifies other plan holders with similar and complementary plans and objectives, and supports regular collaboration focused on integrating with those plans to optimize achievement of individual and collective goals and objectives in an incident.

Lesson: **6 - Additional Resources and Summary**

Lesson Overview

As you learned in the first lesson, the National Response Framework is a compendium of resources, not just a single document. The purposes of this lesson are to:

- Introduce you to the additional resources available to support the implementation of the National Response Framework.

- Summarize the key points presented in this course.

National Response Framework Components

Core Document: The core document presents:

- An **Introduction** to the doctrine that guides our national response.

- **Roles and Responsibilities** including who is involved with emergency management activities at the local, tribal, State, and Federal levels and with the private sector and nongovernmental organizations.

- **Response Actions** that describe what we as a Nation collectively do to respond to incidents.

- **Response Organization** specifying how we as a Nation are organized to implement response actions.

- **Planning** requirements to achieve an effective national response to any incident that occurs.

Annexes

- The **Emergency Support Function Annexes** group Federal resources and capabilities into functional areas that are most frequently needed in a national response (e.g., Transportation, Firefighting, Mass Care).

- The **Support Annexes** describe essential supporting aspects that are common to all incidents (e.g., Financial Management,

Volunteer and Donations Management, Private-Sector Coordination). The actions described in the Support Annexes are not limited to particular types of events, but are overarching in nature and applicable to nearly every type of incident. In addition, they may support several ESFs.

- The **Incident Annexes** address the unique aspects of how we respond to seven broad incident categories (e.g., Biological, Nuclear/Radiological, Mass Evacuation). The overarching nature of functions described in these annexes frequently involves either support to or cooperation of all Federal departments and agencies involved in incident management efforts to ensure seamless integration of and transitions between preparedness, prevention, response, recovery, and mitigation activities.

Partner Guides and Overview Document

The National Response Framework is augmented by the following additional resources:

- **Partner Guides** provide ready references describing key roles and actions for local, tribal, State, Federal, and private-sector response partners.

- The **Overview Document** presents a summary of the process, roles, and responsibilities for requesting and providing all forms of Federal assistance.

NRF Resource Center

The NRF Resource Center includes:

- Authorities

- Overviews of Federal Assistance

- Glossary/Acronyms

- Additional Resources

- Briefings and Training

The Resource Center may be accessed at www.fema.gov/NRF. Be sure to sign up for email updates.

National Response Framework Summary

In this course, you've learned that the National Response Framework:

- Presents the guiding principles that enable all response partners to prepare for and provide a unified national response to disasters and emergencies – from the smallest incident to the largest catastrophe.

- Defines the key principles, roles, and structures that organize the way we respond as a Nation.

- Describes how communities, tribes, States, the Federal Government, and private-sector and nongovernmental partners apply these principles for a coordinated, effective national response.

Key Concept: The Framework fosters response partnerships at all levels of government, and with nongovernmental organizations and the private sector.

Response Doctrine

The Framework establishes a response vision through the following key principles:

- **Engaged Partnership.** Leaders at all levels must communicate and actively support engaged partnerships by developing shared goals and aligning capabilities so that no one is overwhelmed in times of crisis.

- **Tiered Response.** Incidents must be managed at the lowest possible jurisdictional level and supported by additional capabilities when needed.

- **Scalable, Flexible, and Adaptable Operational Capabilities.** As incidents change in size, scope, and complexity, the response must adapt to meet requirements. Given its flexibility and scalability, the National Response Framework is always in effect and elements can be implemented at any level and at any time.

- **Unity of Effort Through Unified Command.** Effective unified command is indispensable to response activities and requires a clear understanding of the roles and responsibilities of each participating organization.

- **Readiness To Act.** Effective response requires readiness to act balanced with an understanding of risk. From individuals, households, and communities to local, tribal, State, and Federal governments, national response depends on the instinct and ability to act.